W0038490

GENETIC RESOURCES AND THEIR EXPLOITATION
CHICKPEAS, FABA BEANS AND LENTILS

ADVANCES IN AGRICULTURAL BIOTECHNOLOGY

Already published in this series

Genetic Resources and Their Exploitation – Chickpeas, Faba beans and Lentils

edited by

JOHN R. WITCOMBE

FAO/IBPGR
South West Asia Program
ICARDA
Aleppo
Syria

WILLIAM ERSKINE

Lentil Breeder
ICARDA
Aleppo
Syria

1984 **MARTINUS NIJHOFF/DR W. JUNK PUBLISHERS**
a member of the KLUWER ACADEMIC PUBLISHERS GROUP
THE HAGUE / BOSTON / LANCASTER
FOR ICARDA AND IBPGR

Distributors

for the United States and Canada: Kluwer Boston, Inc., 190 Old Derby Street, Hingham, MA 02043, USA

for all other countries: Kluwer Academic Publishers Group, Distribution Center, P.O.Box 322, 3300 AH Dordrecht, The Netherlands

Library of Congress Cataloging in Publication Data

Library of Congress Cataloging in Publication Data
Main entry under title:

Genetic resources and their exploitation.

(Advances in agricultural biotechnology)
Includes index.
1. Legumes--Germplasm resources. 2. Chickpea--Germ-
plasm resources. 3. Faba bean--Gerplasm resources.
4. Lentils--Germplasm resources. I. Whitcombe, John R.
II. Erskine, William. III. Series.
SB177.L45G46 1984   635'.65   84-1461
ISBN-13:978-94-009-6133-3   e-ISBN-13:978-94-009-6131-9
DOI: 10.1007/978-94-009-6131-9

ISBN-13:978-94-009-6133-3

Copyright

© 1984 by Martinus Nijhoff/Dr W. Junk Publishers, The Hague, and ICARDA.
Softcover reprint of the hardcover 1st edition 1984

All rights reserved. No part of this publication may be reproduced, stored in a retrieval system, or transmitted in any form or by any means, mechanical, photocopying, recording, or otherwise, without the prior written permission of the publishers,
Martinus Nijhoff/Dr W. Junk Publishers, P.O. Box 566, 2501 CN The Hague, The Netherlands, and ICARDA

CONTENTS

COLLECTION AND MAINTENANCE OF FOOD LEGUME GENETIC RESOURCES

UTILIZATION OF FOOD LEGUME GENETIC RESOURCES

CHICKPEA

Foreword

Chickpeas, faba beans and lentils are important pulse crops in the Mediterranean region and Middle East, where their high protein seed nutritionally complement cereal grain in the human diet. The by-products of these crops serve as a valuable feed for animals. Thanks to their ability to fix atmospheric nitrogen, the inclusion of these crops in the cropping system helps in the maintenance of the productivity of the soil and reduces the dependence of the farmer on fertilizer nitrogen to realise good yields.

Being the site of original domestication of these legumes, the Fertile Crescent is believed to possess their vast genetic diversity. In order to prevent the erosion of this genetic diversity and to preserve it for posterity, it is necessary that a major effort is made for its expeditious collection, evaluation, documentation and safe storage. The International Center for Agricultural Research in the Dry Areas (ICARDA) being located in the Fertile Crescent has, within its mandate, the responsibility to act as a world centre for the work on the genetic resources of kabuli chickpeas, faba beans and lentils. The International Board of Plant Genetic Resources (IBPGR) has been strongly supporting ICARDA in this important activity.

Realising the size of the task involved and the limited availability of trained personnel in the region, ICARDA and the IBPGR jointly organised a training course at ICARDA on the genetic resources of chickpeas, faba beans and lentils in 1982. The material presented by different scientists in that training course has served as the basis of this book. It is the result of a joint endeavour between ICARDA and the IBPGR. The book provides a continuum between collection and evaluation, on the one hand, and the utilization of genetic resources, on the other hand. The aspect of utilization has, hitherto, received little attention in the genetic resources literature. It is hoped that the book may go someway to bridging this gap.

Mohamed A. Nour
Director General
ICARDA

EDITORS' NOTE

We are grateful to Martinus Nijhoff Publishers for permission to print, with
minor changes, the chapter on 'Genetic Resources of Faba Beans' by
J.R. Witcombe, and an adaptation of the chapter 'The Genetic Improvement of Faba
Bean' by G.C. Hawtin. Both these chapters were published in 'Faba Bean
Improvement' edited by G. Hawtin and C. Webb, 1982.

The chapter on 'Taxonomy, Distribution and Evolution of the Lentil and its
Wild Relatives' by J.I. Cubero has been substantially revised from 'Origin,
Taxonomy and Domestication' which was published in 'Lentils' by ICARDA-CAB. We
are grateful to CAB for permission to reprint, with minor changes, 'Genetic
Resources' which was also published in 'Lentils'.

The Editors wish to acknowledge the help of all those people who assisted in
typing and art work. We would particularly mention Aida Bhattikha, Rania
Barrimo, Sylva Cholokian, Lina Khoury and Abdul Rahman Hawwa.

We are grateful to Mr. P. Neate, of the ICARDA Communications Department, for
carefully checking the final draft of the manuscript. Any errors that remain
are the Editors' responsibility. Mrs Rosie Keatinge kindly prepared the index.

The text was prepared and printed at ICARDA's computer centre. We would like
to thank Mr. K. El-Bizri, Director of ICARDA's computing services, for his
invaluable help. Finally, the editors are grateful to Dr. M.C. Saxena, Leader
of the Food Legume Improvement Program, for his encouragement during the
preparation of this book.

ABBREVIATIONS

ALAD   Arid Lands Agricultural Development Program, Ford Foundation, Beirut, Lebanon.

AUB   American University of Beirut, Lebanon.

CAB   Commonwealth Agricultural Bureaux, Farnham, U.K.

FAO   Food and Agriculture Organization of the United Nations, Rome, Italy.

IBPGR   International Board for Plant Genetic Resources, FAO, Rome, Italy.

ICARDA   International Center for Agricultural Research in the Dry Areas, Aleppo, Syria.

ICRISAT   International Crops Research Institute for the Semi-Arid Tropics, Hyderabad, India.

IDRC   International Development Research Centre, Canada.

IRRI   International Rice Research Institute, Los Banos, Philippines.

PBI   Plant Breeding Institute, U.K.

USDA   United States Department of Agriculture, U.S.A.

WRPIS   Western Regional Plant Introduction Station, U.S.A.

PREFACE

The introduction of high-yielding, new cultivars is rapidly transforming traditional and peasant farm lands into fields of uniform crops. The modern cultivars provide significantly higher yields; they are better adapted and have better disease resistance and help significantly to solve world problems of hunger and malnutrition. Unfortunately, these new varieties, with their narrow genetic base, replace the traditional cultivars (the very genetic diversity which was used for the creation of such new varieties) may be quickly responsible for the total elimination of indigenous ancestors (landraces or primitive cultivars) developed over millenia of cultivation.

Crucial as it is to enhance agricultural productivity to alleviate hunger and malnutrition, it is equally important to conserve and make available plant germplasm to meet the increasing demands of plant scientists. Development must not lose sight of the need to conserve genetic resources.

Concern over the loss of valuable sources of plant germplasm has been widely recognized in the past two decades. Many institutions around the world hold collections of crop germplasm and their wild relatives, which were assembled over the last fifty years or so by exploration and seed exchange. They contain thousands of samples and were made, for the most part, either to help plant breeding and agricultural development, or for studies on the diversity and evolution of crops.

The International Board for Plant Genetic Resources was established in 1974 by the Consultative Group on International Agricultural Research (the parent body of the International Agricultural Research Centers (IARC's) such as ICARDA, IITA and IRRI) to coordinate existing national, regional and international plant genetic resources conservation efforts. The FAO provides the Executive Secretariat and Headquarters for the IBPGR. Financial support for the IBPGR is provided by various governments and agencies.

The IBPGR is developing international collaboration among the voluntary members of a global network of genetic resources centres, whose activities will serve to safeguard and make freely available, for crop improvement purposes, the genetic variability of major food crops and other plants of economic importance. The IBPGR is promoting a greater awareness throughout the world of the urgency and need for genetic resources activities. Since the IBPGR was established eight years ago, a great deal of progress has been made towards the organisation of a global network of crop genetic resources centres. At the same time, a large number of exploration and collecting missions have been carried out in a wide range of eco-geographical areas throughout the world.

The Board regards its role as to serve essentially as a catalytic agent encouraging and coordinating global efforts, financing essential activities, e.g. collection, conservation, characterization and preliminary evaluation, documentation and utilization of plant genetic resources. The IBPGR does not pay for all the activities which it coordinates, rather it provides "pump-priming" funds for the initiation of important activities that are expected to become self-supporting.

The Board's tasks also include:
- the establishment of standards, methods and procedures for exploration.
- the evaluation and conservation of genetic stocks of both seeds and vegetative material.
- the establishment of a network of replicated storage centres.
- the promotion of training.

As a part of its strategy, the IBPGR has defined priorities for action on crops and regions. These priorities have been used as guidelines for the initiation and support of programs by the IBPGR. These crops will, of course, continue to receive the Board's attention but, in addition, the IBPGR is adding a few new major crops to its workload each year. During 1982, the IBPGR refined its previously agreed priorities among crops and the list now includes over 50 first priority crops which require action, at least in one particular region, to safeguard germplasm for future use.

To date, much of the Board's efforts have been directed to the major food crops, and have included close collaboration with the sister International Agricultural Research Centers of the CGIAR. In the last two years the work of the IBPGR has taken in tropical vegetables, tropical fruits and a few non-food crops but, of course, these receive less attention than the major food crops. The Board has recently agreed to pay more attention to the conservation of clonally propagated crops.

The Board is also concerned to promote the documentation of important germplasm collections and its lists of internationally acceptable descriptors for various crops. To date, the IBPGR has published descriptor lists for 26 crops or groups of crops. The IBPGR is also compiling information on genetic resources collections of major crops and publishing this in the form of directories. The Board is also making efforts to ensure that information on characterization and preliminary evaluation is recorded and included in data storage systems, along with the collection data. This aspect cannot be stressed too strongly. The Board continues to assist centres in its network in cleaning-up data, putting these into machine-readable form and then into data storage and retrieval systems.

I now turn to training or human resource development. In order that genetic resources activities may be effectively carried out by the various institutions and national programs participating in the global network of the IBPGR, personnel with adequate technical skills are required. There is a lack of trained manpower in many of the developing countries and the strategy of the IBPGR - from its inception - has been to support an international training course in conservation and utilization of plant genetic resources at the University of Birmingham, UK, with IBPGR support being used solely for the benefit of third world participants.

In addition, a number of practical short technical training courses are provided by the IBPGR, primarily in developing countries. These usually deal with basic collection and conservation strategy or special topics needed to move regional cooperative programs forward. The 1982 training course on food legume genetic resources, from which this book was developed, was the second in a series organized jointly by the IBPGR and ICARDA. In 1980, ICARDA organized a training course on wheat and barley germplasm - with financial support from the IBPGR.

Needless to say, legumes rate second only to cereals as a source of human and animal food. Legumes are found in all the continents under varied agro-ecological situations. Food legumes are important sources of nutrients and provide supplementary protein, calories, vitamins and minerals to diets based on cereal and/or starchy foods. Some legumes, such as soybean, groundnut and winged bean, are also rich in oil.

The cultivation of food legumes is ancient, and their centres of origin or diversity are correlated with the development of ancient human societies and their agricultural needs. These remain as a major food in Latin America (especially _Phaseolus_ beans), in southwest and central Asia and the Indian sub-continent (chickpea, lentil, pigeonpea, mungbean), East Asia (soybean) and Africa (cowpea). Any increase in the production levels of these crops is helpful in alleviating the protein deficiency of developing countries. Unfortunately, the yield of many food legumes remains low. As a result, part of the land once used for these crops has gradually been planted with improved varieties of cereal crops.

Four IARCs are concerned with food legumes and their research programs deal with the improvement of quality and quantity of _Phaseolus_ beans (CIAT), lentil, kabuli chickpea and faba bean (ICARDA), pigeonpea, chickpea and groundnut (ICRISAT), and cowpea and soyabean (IITA). The success of any crop improvement depends greatly on the collection and exploitation of genetic variability. From the outset the IBPGR has regarded action on food legume crop germplasm collection and conservation as important and has continued to receive excellent cooperation in this endeavour from the sister CGIAR institutes. An Advisory Committee on _Phaseolus_ beans germplasm, co-sponsored by CIAT, was established in 1976; collection of _Arachis_ germplasm has been carried out since 1976 and an _Ad_

Hoc Working Group on Groundnut Germplasm was convened jointly with ICRISAT in 1979.   As part of the IBPGR on-going activities in Africa, the Board collaborates closely with the Genetic Resources Units of ICRISAT and IITA in their collection and conservation efforts.   The IBPGR has also made funds available to ICARDA and IITA for their storage facilities, and to CIAT and ICRISAT in their germplasm collection efforts.   The IBPGR also convened an Ad Hoc Working Group on Vigna species in September 1981 and at its recent meeting the Board agreed to take action on various Vigna species germplasm.   Besides these crops the IBPGR is also supporting collection and conservation of various other food legume crops and their wild relatives, such as winged bean, lupins and peas.

N. Murthi Anishetty
IBPGR, Rome, Italy

1. COLLECTION AND INITIAL PROCESSING
OF FOOD LEGUME GERMPLASM

J.R. Witcombe
FAO/IBPGR, South West Asia Program

Introduction

This chapter is intended to be a practical guide for the collection of food
legume germplasm (or any grain crop for that matter) and its initial
documentation. Although it is assumed that the documentation is being handled
manually, the methods recommended are very suitable for subsequent
computerisation.

There is a need to collect the local varieties of food legumes, either
because they are being replaced by introduced cultivars or because they are in
danger of being replaced. The replacement of local cultivars by exotic ones has
been termed genetic erosion. However, even in the absence of genetic erosion,
or its threat, collections are still assembled for scientific investigation and
for a source of selections in breeding programs. When such collections are
assembled they are still in danger of loss. For example, the faba bean
collection studied by Muratova and the lentil collection studied by Barulina
(both students of N.I. Vavilov) are certainly no longer intact. Once
collections have been made they need to be documented and carefully stored and
duplicated to minimize the risk of loss.

Collection strategy

Patterns of variation and sampling technique

The patterns of variability found on a world-wide basis were first described
by Vavilov, who formulated the concept of 'Centres of Diversity' (Vavilov,
1926). Patterns of variation have been reviewed by, amongst others, Frankel and
Soulé (1981), and Murphy and Witcombe (1982). All work on variation in
germplasm material indicates that variability is not uniformly distributed on a
macro-geographic scale. However, the lack of uniformity tends neither to be
very marked nor to follow the pattern of Vavilovian centres of diversity.

Murphy and Witcombe (1982) and Witcombe and Gilani (1979) have shown that, at
the very least in wheat and barley, the theory of 'Centres of Diversity' is
restricted to the characters examined by Vavilov. These characters can be
regarded as 'qualitative' ones since they are highly heritable, conspicuous and
easily determined by collectors in the field. In contrast, quantitative
variation does not follow Vavilovian patterns. Thus, on both a macro-geographic

J.R. Witcombe and W. Erskine (eds.) Genetic Resources and Their Exploitation - Chickpeas, Faba
beans and Lentils. ISBN-13:978-94-009-6133-3 (PB)
©1984, Martinus Nijhoff/Dr W. Junk Publishers for ICARDA and IBPGR.

and local scale, samples should be taken over as wide an area as possible and collection should not be concentrated in Vavilovian centres.

Certain practical guidelines have to be followed in an attempt to sample fully the range of variability, whilst avoiding the inefficiency of collecting numerous identical samples.

Examination of populations of cultivated wheat and barley from the Himalayas has shown that variation is distributed according to regions (Witcombe and Rao, 1976). The varieties of any one region tend to be alike, and the closer together any two collection sites are, the more likely it is that the samples from them are similar. Although there is little other work which has examined this problem, restricted seed interchange will mean that, in many crops, varieties from the same village or closely adjacent villages will be alike. Therefore, it is obviously a useful strategy to avoid collecting from sites that are very close together. As a practical rule, collections should not be made from sites that are less than 10 km apart, unless:

- It can be seen that the varieties are morphologically different.
- There has been a marked change in altitude or cropping system.
- A formidable barrier to communication has been crossed, such as a mountain pass or a river which cannot be easily crossed.
- The local people are ethnically different from those in the previous collection site.

The '10 km rule' can be reduced in the case of very difficult terrain, where it is essential to travel on foot, and increased in regions where there are good roads. Also, in the initial stages of a collecting program, a greater distance than 10 km can be used if the collector is confident of being able to return in a subsequent season. The collected material can then be evaluated to determine the amount of variability present, and further collections can be concentrated where the crop has the greatest variability.

The above applies to cultivated crops where the seed can be transported from village to village by the local farmers, and where there is the possibility of collecting exactly the same variety in adjacent villages. Different considerations apply for wild relatives of food legumes as they grow in the same site each year, with a limited genetic interchange between populations by the movement of seed and pollen. A considerable body of evidence, from the study of adaptation and the formation of ecotypes, shows that the genetic variability between wild populations is related to differences in their environments (e.g. Bradshaw et al., 1965). Sampling should exploit this fact and be more frequent in areas where there are many different environments.

Since wild populations can display large genetical differences over very short distances (even a few meters where there is a considerable environmental change) they can be sampled at much closer distances than in the case of cultivated crops.

In general, for both cultivated and wild species, sampling should be over as many different environments and regions as possible, and attempts be made to cover the largest area in the time available.

It is a good plan to collect away from the major routes since any previous collections are more likely to have been made along them. This becomes particularly important where the introduction of advanced cultivars has started, since this usually begins in the most accessible regions close to the major roads. It is vital that these regions away from the roads are explored, even if it is necessary to leave the vehicle and travel on foot or hired animals.

Finally, in planning any sampling strategy, it is essential to keep careful records, which should include maps, as to where collections have been made and to use these records when in the field. If this is not done then in subsequent seasons collections may inadvertantly be repeated, or made from locations extremely close to existing collecting sites.

Avoid introduced cultivars

It is obviously important to avoid collecting introduced, improved cultivars. To do so it is helpful to interview the local farmers as they often will be able to tell you whether the crop has been introduced. Such information will usually be reliable. However, information that the variety is local cannot always be trusted, because a farmer often forgets the past introduction of a variety that he has grown for many years. In food legumes the problem of collecting introduced cultivars will often be a minor one, because of the current lack of improved cultivars, but it should not be disregarded.

The simplest way of avoiding the collection of introduced varieties is to know the crop and to collect during the harvest season. Although introduced cultivars can usually be eliminated after initial evaluation, this is much more time-consuming and more complicated than not collecting them in the first instance.

Collecting efficiency/sample size

There are two simple ways to save considerable time in the field. The first is to plan the itinerary to avoid, as far as is possible, retracing the route; this may involve camping and staying in places where accommodation is less than ideal. The second is to make the greatest use of the time that has been spent at a site (stopping, identifying and describing the site, and perhaps finding and interviewing a local farmer) by collecting all of the crops and related species that are available.

An important factor in determining efficiency is the time spent in collecting each sample, for the larger the sample the greater the time taken. However, an important exception is a collection from a farmer's grain store or threshing place when (apart from perhaps extra time spent in negotiations) a large sample takes no longer to collect than a small one. If such seed is available then as much as one kilogram can be collected.

When collections are made in the field, a balance has to be achieved between a large number of small samples or a smaller number of larger samples. There is no theoretical answer (although attempts have been made to derive one) as to the optimum size of each sample, and in the case of cultivated crops, pods or seed from about 100 to 200 plants seems to be a sensible amount. This provides an adequate sample, yields a reasonable amount of seed and does not take an excessive time to collect.

It is very useful to collect sufficient quantities of seed to be able to send sub-samples before multiplication. The danger of loss is minimised and the material is distributed more rapidly. In my opinion, optimum sample size is determined by this simple, practical consideration rather than by theoretical population genetics.

With wild species, it is difficult to specify any quantity since, depending on the species, its maturity, and its local abundance, the time to collect a set quantity of grain can vary enormously. The timing of the collection is very critical since harvesting needs to be done at the near-ripe, or just-ripe, stage before the pods dehisce. However, if possible, it is again best to collect sufficient seed to allow for distribution of a sub-sample before multiplication.

The collection record

An example of a collection record form is shown in Appendix 1. To save time in the field the form requires a minimum quantity of data and excludes soil description and topography, since the former is very time consuming to do properly, and the latter is subjective and not essential. Although aspect and slope are included in the form, they are only relevant to wild and forage species which, in contrast to annual crops, are known to remain in the same site over a number of years.

The forms can be bound in hard-backed, loose-leaf folders in lots of 50 with the site number of each 'Field Book' beginning at 1, 51, 101 and so on. This form uses only one page per site and is more compact and rapid than the conventional method of using one page per collection (Appendix 2). The use of a one page per site method should not preclude the use of collector's numbers. These numbers are, for example, made up of the site number and the sample number. Thus we have collector's number 123-1, which is sample one from site 123. The collector's number could also be prefixed by the team, to give e.g. JWWE 123-1. Otherwise the collector's number is the same as the site number.

In the field

The collection bag should ideally be labelled both within and on the outside of the bag. The label inside can be a tear-off tag which can accompany the sample when it is cleaned, threshed, or placed in its storage container. It is obviously an advantage to have a label on the outside of the bag as this facilitates the initial sorting of the samples.

The most essential information is the precise identification of the site. Unless this is done, the entire strategy of avoiding collection from sites that are very close together cannot be followed. If the location site is not obvious from the map, then the names of adjacent villages and kilometer readings of the vehicle (if applicable) at known places before and after the site are essential data.

The reverse of the form can be used to enter any additional information.

On return to the laboratory

To save time while in the field, the following items can be completed after returning to the laboratory:

- Latitude and longitude (it is simply not practical to record these in the field).
- The Latin names of crops.
- Expedition/Team (a rubber stamp can be used).

Entering the collected material into the gene bank

Once the material has been collected and brought back to the laboratory the first task is to document the material and store it in appropriately labelled containers. This process is done in the following six steps:

i.   The collection record forms are completed.

ii. A helpful aid is to make a list of collections in order of field (i.e. collector's) number. When sorting the material, it is easier to use a list than to leaf though the pages of the field book.

iii. The external labels on the collecting bags are used to arrange the samples crop by crop according to the accession numbering system in use.

iv. Accession numbers can now be assigned to each crop in turn. From this stage on, the accession numbers are the means of identifying samples. They are entered in the field book and in the list in (ii) above.

v.   The material is threshed, cleaned and dried as necessary, and placed in storage containers labelled with accession numbers. Each crop is again treated in turn.

vi. Each accession is recorded in the accession book by entering at least its accession number, site number and field (collector's) number. As an example, a page from an accession record book, used at ICARDA, is shown in Appendix 3.

The accession book records the essential data for each accession and also cross refers to other data files. The site number refers to the site data list (see below) and the field book.

The material has now been entered in the gene bank. What happens next to any sample will depend on its size. If very large quantities are available, a large sub-sample can be used as a base collection for long-term storage, smaller sub-samples can be distributed to other institutions, and sub-samples prepared for sowing for preliminary evaluation. Alternatively, if there is only a small quantity, the seed has to be multiplied before any distribution can take place.

Handling site data

Although the site data form an integral part of the documentation of the accessions, it is more convenient to deal with them as a separate task after the more urgent work of preparing the collected material for storage has been completed. Only the first step in the list below has to be made before the accessions are entered into the gene bank (and normally this step will not be necessary). The rest of the work on site data can be left until later.

i.   If several teams have been collecting in the field at the same time, the site and collector's numbers will have to be revised if different teams have used the same ones. This problem should be avoided by allocating different sets of numbers to the various teams - in any case the numbers in the field books are easily changed so that no two sites share the same number.

ii.   By means of the data in the field books, and maps of as large a scale as possible (1:250,000 is usually the best available), the precise location of each collecting site is determined. When site locations are found on the map, these positions are marked with a site number (a yellow felt tip pen or textmarker pen is clearly visible without obscuring any detail on the map).

iii. A list of site data can then be made.

iv.   A master map is prepared (scale 1:1,000,000) in which every site is marked. This site map is used, in conjunction with the accession record lists, to prepare separate maps which show the distribution of each crop collected throughout the region.

The above mentioned method gives each collection site a number which is used to identify the site on the map and in the list of site data. Each site can then be labelled on the map with a single number, instead of with sets of collector's numbers. The data of any one site are only written once in the site data list. The site number is entered, as a reference to this list, in the accession record to avoid writing complete site data against every accession. This has a particular advantage when several accessions have been collected from the same site, as the site data does not have to be repeated.

Since a site data list can be typed on a few pages and contain all the collection sites of a gene bank, a copy of this list can be sent whenever accessions are distributed. A list of accessions for distribution purposes then requires only three entries for each accession; (1) the accession number (2) the site number and (3) the collector's number. If the accession list has more than one species, then the name of the species will also be required.

Summary

After the collected material has been entered into the gene bank and all site data have been dealt with, the following have been completed:

i.   The collection record forms- 'Field books'.
ii.   An accession record list in order of collector's number (an aid to sorting the material).
iii.   An accession book for each crop (which refers to i and iv).
iv.   A list of site data.
v.   Maps of sites and crops.

Alternative method

Site numbers are not used, but complete site data are entered against every accession in the accession record lists and distribution lists. The sites are labelled on the maps with field numbers. This is a more conventional method.

References

Bradshaw, A.D., McNeilly, T.S. and Gregory, R.P.G. 1965. Industrialization, evolution, and the development of heavy metal tolerance in plants. British Ecological Society Symposium 5:327-343.

Frankel, O.H. and Soulé, M.E. 1981. The genetic diversity of plants used by man. Pages 175-223 in Conservation and Evolution. Cambridge University Press, U.K.

Murphy, P.J. and Witcombe, J.R. 1982. Variation in Himalayan barley and the concept of centres of diversity. Pages 26-36 in Barley Genetics IV. Edinburgh University Press.

Vavilov, N.I. 1926. Centres of origin of cultivated plants. Bulletin of Applied Botany and Plant Breeding, Leningrad 16: 139 - 248.

Witcombe, J.R. and Gilani, M.M. 1979. Variation in cereals from the Himalayas and the optimum strategy for sampling plant germplasm. Journal of Applied Ecology 16: 633 - 640.

Witcombe, J.R. and Rao, A.R. 1976. The genecology of wheat in a Nepalese centre of diversity. Journal of Applied Ecology 13: 915 - 924.

SITE DATA

EXPEDITION	TEAM	DATE	SITE NO.	PHOTO

PROVINCE —————————— VILLAGE NAME —————————— KM READING ——————————
PRECISE LOCALITY ——————————————————————————————

MAP SHEET	LATITUDE N S	LONGITUDE E W	ALTITUDE	ASPECT	SLOPE

ADDITIONAL INFORMATION ——————————————————————————

SAMPLES

1 COLLECTORS NO. | GENUS | SPECIES/LOCAL NAME

WILD/CULT. IRR/DRY. SUMM/WINT. SOWN HARV. THRESHED/UNTH. ORIGIN* SAMPLE SIZE
1 SINGLE PLANT 3 LARGE POP.
2 SMALL POP.

2 COLLECTORS NO. | GENUS | SPECIES/LOCAL NAME

WILD/CULT. IRR/DRY. SUMM/WINT. SOWN HARV. THRESHED/UNTH. ORIGIN* SAMPLE SIZE
1 SINGLE PLANT 3 LARGE POP.
2 SMALL POP.

3 COLLECTORS NO. | GENUS | SPECIES/LOCAL NAME

WILD/CULT. IRR/DRY. SUMM/WINT. SOWN HARV. THRESHED/UNTH. ORIGIN* SAMPLE SIZE
1 SINGLE PLANT 3 LARGE POP.
2 SMALL POP.

4 COLLECTORS NO. | GENUS | SPECIES/LOCAL NAME

WILD/CULT. IRR/DRY. SUMM/WINT. SOWN HARV. THRESHED/UNTH. ORIGIN* SAMPLE SIZE
1 SINGLE PLANT 3 LARGE POP.
2 SMALL POP.

5 COLLECTORS NO. | GENUS | SPECIES/LOCAL NAME

WILD/CULT. IRR/DRY. SUMM/WINT. SOWN HARV. THRESHED/UNTH. ORIGIN* SAMPLE SIZE
1 SINGLE PLANT 3 LARGE POP.
2 SMALL POP.

6 COLLECTORS NO. | GENUS | SPECIES/LOCAL NAME

WILD/CULT. IRR/DRY. SUMM/WINT. SOWN HARV. THRESHED/UNTH. ORIGIN* SAMPLE SIZE
1 SINGLE PLANT 3 LARGE POP.
2 SMALL POP.

* 1: FIELD , 2: THRESHING PLACE , 3: FARM STORAGE , 4: EXPERIMENTAL STATION , 5: OTHER.

IBPGR COLLECTION FORM (GENERAL)

Descriptors in this column **MUST** be filled in

GENUS: _____

SPECIES: _____

SUBSPECIES: _____

COLLECTOR'S NUMBER: _____

COLLECTING INSTITUTE: _____

DATE OF COLLECTION: _____

COUNTRY OF COLLECTION: _____

PROVINCE/STATE: _____

LOCATION OF COLLECTION SITE

    nearest town/village: _____

    distance (in Km): _____

    direction: _____

LATITUDE OF SITE: _____ N   S ___

LONGITUDE OF SITE: _____ E.   W ___

ALTITUDE OF SITE: _____ (m) _____

COLLECTION SOURCE (circle one)

wild	1	village market	5
farmland	2	commercial market	6
farmstore	3	institute	7
backyard	4	other (specify)	8

STATUS OF SAMPLE (circle one)

wild	1	primitive cultivar/landrace	4
weedy	2	advanced cultivar (bred)	5
breeder's line	3	other (specify)	6

LOCAL NAME: _____

NUMBER OF PLANTS SAMPLED: _____

PHOTOGRAPH (circle one):   yes   no

    Photo number: _____

TYPE OF SAMPLE (circle one)

    vegetative 1   seed 2   both 3

HERBARIUM SAMPLE (circle one): yes   no

QUANTITY OF MATERIAL (number of seeds or plant
    samples): _____

Descriptors in this column **SHOULD** be filled in

CULTURAL PRACTICES:

shifting (circle one)	yes	no
irrigated (circle one)	yes	no
transplanted (circle one)	yes	no
terraced (circle one)	yes	no

SOWING MONTH: _____

HARVEST MONTH: _____

USAGE (specify): _____
_____

DISEASES AND PESTS (specify): _____
_____

ASSOCIATED WILD AND WEEDY SPECIES AND CROPS
(specify): _____
_____

TOPOGRAPHY (circle one)

swamp	1
flood plain	2
plain level	3
undulating	4
hilly	5
mountainous	6
other (specify)	7

SITE (circle one)		STONINESS (circle one)	
level	1	none	1
slope	2	low	2
summit	3	medium	3
depression	4	rocky	4

SOIL TEXTURE (circle one)		DRAINAGE (circle one)	
sand	1	poor	1
loam	2	moderate	2
clay	3	good	3
silt	4	excessive	4
highly organic	5		

OTHER OBSERVATIONS: _____
_____
_____
_____

APPENDIX 3

		Accession number	Species/subspecies (code)	Donor Organization	Donor Number	Collecting (Breeding) Organization	Collector	Collection Number (Variety/Pedigree)	Date of Collection	Country of Collection (Where Bred)	Province/Town	Altitude	Latitude	Longitude	Other site data/ Remarks	Date received by ICARDA
				Donor				Collection (or breeding) data								

2. SEED STORAGE, VIABILITY AND REJUVENATION

L.J.G. van der Maesen
ICRISAT, Hyderabad, India

Purpose of seed storage

Agricultural seeds often need to be retained for use for more than one season or year, and losses of viability occur if protection against heat, moisture, and pests is not provided. Even the most primitive agriculturist in remote history learned to protect his seeds from deterioration by drying them and by using closed containers to exclude moisture and pests, and any trade in seed is impossible without storage.

In the case of genetic resources, seed needs to be stored for a long period to be available to plant breeders in their quest for particular genes. There are two kinds of collections of genetic resources :

i.   Base collections for long-term storage as a precaution against loss.

ii. Active collections under medium-term storage used for seed distribution, evaluation, and multiplication.

Principles of seed storage

Seed physiology is a rare scientific specialization and detailed comparative data on the subject are few and rather scattered; but the general principles are fortunately simple. Roberts (1972) and Justice and Bass (1978) have produced excellent handbooks. Barton (1961) is authoritative on the longevity of seeds.

Variation among species

Plant species can be divided into three seed storage groups.

i.   Orthodox seeds are those which can be stored without any problem for long periods in dry and cool conditions, and they include chickpea, faba bean and lentil, most cereals, and many other agricultural crops. For example, barley seeds sealed in glass tubes and kept at room temperature had a germination of 12% after 123 years.

ii. Recalcitrant seeds have a short life span and cannot be dried since they do not tolerate a loss of moisture. This group includes many tropical crops, such as cocoa, sugarcane, coconut, citrus, rubber, and tea. Their seed characteristics are much less well known than those of group i. Conservation in vitro, as meristem cultures, may be a solution to the storage problems of this group.

J.R. Witcombe and W. Erskine (eds.) Genetic Resources and Their Exploitation - Chickpeas, Faba beans and Lentils. ISBN-13:978-94-009-6133-3 (PB)
©1984, Martinus Nijhoff/Dr W. Junk Publishers for ICARDA and IBPGR.

iii. Seeds with unknown, or little known, duration of viability under storage. This group includes most wild species.

Natural selection, according to scientists at Fort Collins, is still much more important as a cause of variability than alterations occurring in storage; any mutant cells that are formed will mostly be swamped by normal cells as they are usually weaker or even lethal. The Leguminosae include hard-seeded species, the seeds of which may be long lived. For example, Trifolium seeds have been reported to survive for 100 years; Goodia for 105 years; Cassia for 158 years; and Albizia for 147 years. Faba beans, lentils, and chickpeas, that have been selected for edibility and cookability, do not have very hard seed coats and do not remain viable for so long.

Variation among cultivars

Variation among cultivars is often significant. For example, kabuli chickpeas survive shorter periods of storage than desi chickpeas, the latter having a thicker, harder seed coat. Inheritance studies of seed longevity of pulses have not been reported.

Condition of the seeds

Ideally, stored seeds should be fully-matured, of normal size, uninjured, without storage pests and micro-organisms and unaffected by extreme temperatures and moisture conditions during filling and ripening. Initial high viability confers better resistance than low viability to unfavourable storage conditions. The weather at harvest has an important influence on viability. Sometimes, therefore, plants sown for germplasm stocks are grown in off-seasons. For example, sorghum and millet in India produce better seeds in winter because they mature under drier conditions than in the main (rainy) cropping season.

Hand and mechanical threshing should be carried out with care because damaged seeds lose viability more quickly than undamaged seeds. Seed structure, the ease of removal of the seed from the pod, the moisture content of fruits, and the stage of maturity can influence seed damage.

Seed dormancy

Dormancy is found in several crop species. But not all seeds in one lot are necessarily dormant. Seed dormancy is often a reason for plant quarantine to reject samples as they are suspected to be dead. In wild plants dormancy is more common since cultivated plants have lost this attribute. Cicer montbretii, for instance, is known to survive in soil for three years prior to germination. This is probably because it has a hard seed coat, preventing the exchange of water and gas; but even scarified seeds show dormancy. Bracts, glumes, pericarp, and membranes are other seed structures inducing dormancy, whilst embryo physiology and germination inhibitors can also be involved.   For germplasm collection and maintenance purposes, seeds of wild _Cicer_ and _Cajanus_ are routinely scarified to reduce seed dormancy.   Alternatively, seeds of at least 2-3 years old should be used.   In cold storage dormancy will decrease more slowly than in ordinary temperatures.   Genetic differences in dormancy have been found in lentils (Tosun et al., 1980).

Moisture content

This is one of the most important factors affecting seed longevity.   In general, the drier the seed the greater its longevity.

Food legumes harvested in a dry climate usually have a moisture content of 8-11% in the seed.   At these moisture levels these is no risk of freezing damage, if the seeds are placed in air-tight containers in a cold store. However, if the seed is dried to 5 or 6% moisture content then (assuming a reduction of about four percent in moisture content and a doubling of seed longevity for each one percent decrease in moisture content) an increase in longevity of 2^4 is obtained.   Below 2% moisture content desiccation injury occurs at most temperatures.

High relative humidity causes the seed moisture content to be high.   So, if containers are not vapour and air-proof a cooled store room needs to be dehumidified.   The techniques used to reduce the moisture content of seeds for storage in gene banks are discussed by Witcombe (Chapter 3).

Studies relating to air and seed moisture content are available for many species.   During the monsoon in India, chickpea seeds of kabuli cultivar L550 increased in moisture content from 7 to 12%, while seed of desi cultivars increased by only 1% (observation at ICRISAT).

Temperature

As a 'rule of thumb', between 0 and 50°C seed life is halved for each 5°C and 1% increase in moisture content (Harrington, 1970).   If degrees Farenheit plus % RH total 100 or less, conditions are considered favourable for longevity (another rule of thumb suggested by James, see Roberts, 1972).   Roberts (1972) developed viability nomographs or nomograms giving the time taken for viability to fall to any given level at any given temperature and moisture level.

If variability levels are plotted on a probability (log) scale, the resultant graphs are straight lines instead of normal distribution curves.

The IBPGR has established two standards relating to seed storage:

i.   Preferred standards: -18°C or less; seeds in airtight containers at a seed moisture content of 5 ±1% wet weight (a requirement for long-term storage).

ii.   Acceptable standards: $+5^{\circ}$C or less in airtight containers at seed moisture levels of 5-7%; or $5^{o}C$ in unsealed containers, with a controlled relative humidity (RH) of 20%.

A level of 20% RH is not expensive to maintain in a well designed store (Witcombe, Chapter 3) and higher levels, of 30-40% RH, can easily be achieved.

The freezing of dry seeds considerably increases longevity, perhaps by centuries.   In Fort Collins (USA) trials of storage in liquid nitrogen ($-197^{o}C$) have so far resulted in no seed damage, although experiments have run for only a few years. Life processes are virtually stopped and also, presumably, the mutations that occur in storage. Repeated freezing and thawing, however, may damage seeds.

Vacuum and gas storage

Results of research on this subject are contradictory. At high moisture levels, oxygen shortens the viability of barley, pea, and other seeds. Living seeds use up oxygen in sealed containers, so well-filled closed containers are recommended. For most situations the added expense of using storage atmospheres other than air appears unnecessary. Such atmospheres may be advantageous for very long-term storage, but samples cannot then be easily taken. The control of seed moisture and temperature are of much greater importance than that of the gas medium in seed storage.

Chemicals and pests

Chemicals applied to control fungi, bacteria, viruses, pests and rodents may affect germination and longevity. Dry seeds stored cool are little affected until germination, as the life processes of the parasites also cease. The use of naphthalene balls has proved effective and convenient (except for the odour and possible health damage to workers continually exposed to vapour) to avert insect damage, but in cool storage this is unnecessary. There is no established evidence that exposing seeds to napthalene causes loss of viability.

Mercuric chloride, liquid organic mercurials, and hot water (used in disease control) all reduce seed longevity. Fungicidal dusts usually cause no injury, but there are many exceptions. At low temperatures injury from chemicals is reduced. Fumigants vary in their influence on seed germination. For example, organic mercury reduces the viability of eggplant seeds, but not of carrot, pea, pepper, and tomato. Methyl bromide reduces the vigour and viability of seedlings of several crops, especially when applied in high temperature or moisture regimes.

Fungi do not grow on seeds with a moisture content of less than 12%, and seeds containing less than 9% moisture are rarely attacked by insects.

Sample size

The number of seeds per sample should be at least 12000 for long-term storage. An accession of an inbred plant is sufficiently represented with 4000 seeds. In practice often the container size, or row length in the field, decides how many seeds are kept. The large-seeded grain legumes, in particular faba bean cultivars, require larger than average storage space. Opinions still differ on how many of such seeds should be kept.

Germination tests

After harvesting and drying, the seeds should be subjected to an initial germination test. Methods for germination tests are given in the International Rules for Seed Testing (ISTA, 1966), and they prescribe the optimum method and temperature for many species. Cultivar differences exist, and must be correlated to seed moisture content, determined by weight change on drying or electric moisture meters. Grain legumes are tested in sand or petri dishes, or between paper towels at favourable temperatures. Various equipment is used to facilitate the sampling, counting, and germinating of seeds (ISTA, 1966; Justice and Bass, 1978). A convenient method is to germinate the seeds on moist filter paper in 9-cm petri dishes at 20-30°C. The data should be recorded on standard forms and an example is shown in Appendix 4. Chemical tests (e.g., with tetrazolium) kill the seeds, but are very useful in research to pinpoint dead and living areas in the seeds.

In long-term cold storage 2-4 samples of 50-100 seeds should be taken every 5-10 years to measure their viability, although it may well be found that the viability period is very long. Sampling reduces diversity, both in respect of numbers as well as of populations if the sample is heterogeneous. Sampling should be at random. But it should be noted that even seeds in a homogeneous, homozygous sample will differ in viability.

To economize on seed, it is an excellent practical idea to keep a few samples of seed in large quantities in the cold store under the same conditions as the germplasm material. These samples can be then tested frequently, say yearly, using large samples.

Ellis et al. (1980) have proposed a method of germination testing, 'sequential testing', that uses fewer seeds than the ISTA method (Table 1).

An initial quantity of 40 seeds is tested. Depending on the results of this test the accession is either classified as requiring rejuvenation, or a further test is made . At this second test, and all subsequent tests in the sequence, there are three choices depending on the result of the test.

i.   If the germination is below a certain level the accession requires regeneration and no further testing is necessary.

ii.   If the germination is at an intermediate level, the result is still unclear and a further test is required.

iii. If the germination is high there is no need for further testing and the accession is retained in the cold store.

Table 1. Sequential germination test plan for 85% regeneration standard, testing seeds in groups of 40 (Ellis et al., 1980)

No. of seeds tested	Regenerate if no. of seeds germinated is <=:	Retest accession if no. of seeds germinated between:	Keep in store if no. of seeds germinated is >=:
40	29	30-40	-
80	64	65-75	76
120	100	101-110	111
160	135	136-145	146
200	170	171-180	181
240	205	206-215	216
280	240	241-250	251
320	275	276-285	286
360	310	311-320	321
400	345	346-355	356
440	380	381-390	391
480	415	416-425	426
520	450	451-460	461
560	485	486-495	496
600	520	521-531	532

Storage of pollen

Pollen from a few families (Leguminosae, Primulaceae, Pipaceae) can remain viable for long periods if kept cool and desiccated. Gramineae pollen is short-lived. Ageing factors are mostly the same as for seeds: respiratory substrate exhaustion, enzyme inactivation, desiccation injury, blocking by secondary metabolic products. Pollen storage is complicated and collection of legume pollen is tedious, and is not yet a practical method of storage.

Rejuvenation

Theoretical considerations

Rejuvenation of germplasm is needed when either germination drops below a certain level (50%, 80%) or seed stock falls below an acceptable minimum, which varies with species, breeding system, and opinion. It is wise not to have less than 1000 seeds in stock, but in large-seeded legumes practice often dictates otherwise. Seed increase should preferably be carried out in the area of origin, or one similar to it, in the usual season, or if necessary in the off-season so that ripening takes place in dry weather.

Chickpea and lentil can be grown without isolation, because natural outcrossing is very low (<1%). Faba bean has a rate of outcrossing which varies between 20-40%; hence rejuvenation of faba germplasm must be done in isolation. The methods of isolation in faba bean are discussed in Witcombe (Chapter 12).

An outbreeding species ideally needs to be maintained as a population, however impractical this often is. Such populations contain considerable heterogeneity, which poses problems in characterization and evaluation. The number of accessions or populations could be reduced by merging similar accessions into gene pools. Timothy and Goodman (1979) discourage the formation of gene pools as tools for the maintenance of genetic resources unless they are carefully manipulated into separate usable units. The bulks would contain accessions brought together on the basis of similar characteristics, such as geographical background, or multivariate (cluster) analysis of evaluation data (see trait specific gene-pools in Witcombe, Chapter 12).

Management

In practice the plot size used for the maintenance of germplasm accessions is small. A plot of 2 x 2 m of chickpeas often gives enough seed to fill a bottle with 0.5-1.0 kg capacity. At ICRISAT chickpea is grown in double rows of 4 m, spaced 25 cm apart on the top of ridges spaced 60 or 75 cm apart, but rows or plots on the flat are equally suitable. The recommended row spacing for lentils is 25-30 cm to give a plant population of around 100 plants/m^2. This will ensure a high rate of seed multiplication. Grouping of material of similar maturity is useful, but this is often difficult to plan, especially if the accessions have not been previously evaluated.

Fertilizer application and irrigation are needed to ensure proper stands and satisfactory yields. The amount of irrigation required varies according to available soil moisture and rainfall. Since overdoses may lead to an excessive vegetative growth that reduces yield and makes harvesting cumbersome, or to infestation with root diseases, starter dosage rates of 20 kg of N and 60 kg of P_2O_5/ha are recommended.

Harvesting very small plots must necessarily be done by hand.   Threshing can be done by small threshers, but care must be taken to empty the machine entirely between samples to avoid contamination.   Drying is important.   The drier the samples the less need there is for artificial drying.   Seeds are first put into bags, preferably labelled both inside with a loose tag and outside with an attached label.   In various cleaning operations the loose label can be shifted from tray to tray.

Germplasm rejuvenation (and evaluation) plots should be well protected against pests and diseases.   Accessions not well adapted to the environment under which they are grown may otherwise be lost.   Screening against pests and diseases should always be conducted separately.   Great attention must be paid to avoid mistakes in labelling and the inadvertent mixing of accessions.

Observations

If germplasm has already been sufficiently evaluated, only a few notes such as flower colour, flowering date, and growth habit need to be taken to ensure the identity of the accession.   Identification of off-types is necessary in both lines and populations.   Off-types may be sufficiently interesting to be maintained separately.

Frequency of rejuvenation

The frequency of rejuvenation is often determined more by seed supply requirements than by considerations of the longevity of the seed.   The storage of large amounts of seed is cheaper than frequent grow-outs, and reduces the chance of mistakes.   Errors can be made at all levels, but must be reduced to the minimum.

Duplicate accessions

At ICARDA, ICRISAT, and other international institutes a number of samples is maintained that bear the same accession name ('administrative' duplicates) but differ in regard to seed and plant characteristics.   Commonly, samples have been obtained from different sources under the same name.   All such samples need to be maintained separately unless the description of the original line is known and staff can thus discard the wrongly labelled lines.   Natural selection on a single variety of faba bean grown in different places will tend to result in different populations being formed, because of the different selection pressures on the original heterogeneous variety and outcrossing.

The problem of maintaining real duplicates is more complex. If the size of a collection is manageable, it is better to be cautious and maintain duplicates, unless all descriptors match after several grow-outs. It is difficult to be sure that all genes of two genotypes are similar if the phenotypes are similar. An example is resistant sub-lines, containing genes or alleles for resistance not present in all seeds of an otherwise homogenous population sample. On the other hand, one should never attempt to split a collection into the most refined set of inbreds. For the purpose of germplasm maintenance, populations are preferable. Subsampling, except for useful genes, leads to an explosion of numbers and an unmanageable situation.

References

Barton, L.V. 1961. Seed preservation and longevity. London: Leonard Hill.

Ellis, R.H., Roberts, E.H. and Whitehead, J. 1980. A new, more economic and accurate approach to monitoring the viability of accessions during storage in seed banks. Plant Genetic Resources Newsletter 41: 3-18.

Harrington, J.F. 1970. Seed and pollen storage for conservation of gene resources. Pages 501-521 in Genetic Resources in Plants; Their Exploration and Conservation, IBP Handbook no.11, eds. O.H. Frankel and E. Bennett, Oxford and Edinburgh: Blackwell.

International Seed Testing Association. 1966. International rules for seed testing. Proceedings of the International Seed Testing Association 31: 1-152.

Justice, O.L. and Bass, L.N. 1978. Principles and practices of seed storage. USDA Agricultural Handbook 506. Washington DC: Government Printer.

Roberts, E.H. 1972. Viability of seeds. London: Chapman and Hall.

Timothy, D.H. and Goodman, M.M. 1979. Germplasm preservation: the basis of future feast or famine. Genetic resources of maize: an example. Pages 171-200 in The Plant Seed: Development, Preservation, and Germination, ed. I. Rubenstein, New York and London: Academic Press.

Tosun, O., Eser, D. and Gecit, M.H. 1980. Dormancy in lentils. LENS 7: 42-46.

Appendix 4. Form for germination test data.

GERMINATION TEST LIST

| ACCESSIONS : |
| DATE OF COMMENCEMENT : |
| TEMPERATURE : |
| GERMINATION CONDITIONS : |
| SAMPLE SIZE : |

ACCESSION	HARVEST	NUMBER OF SEEDS GERMINATED				FINAL %
		DATE :	DATE :	DATE :	DATE :	GERM-
NUMBER	YEAR	DAY :	DAY :	DAY :	DAY :	I NATION

3. SEED DRYING AND THE DESIGN AND COSTS
OF COLD STORAGE FACILITIES

J.R. Witcombe

FAO/IBPGR, South West Asia Program

Introduction

In the preceding chapter the key to the successful storage of samples of orthodox seeds such as chickpeas, faba beans and lentils was shown to be the control of the temperature and moisture regime. The 'rules of thumb' are that each reduction of one per cent in seed moisture content, and each reduction in temperature by 5°C doubles the longevity of the seed. The methods of storing seed under these ideal conditions of low moisture content and low temperature are discussed.

Seed drying

Seed is dried by exposing it to air at low relative humidity. This can be done by:

i. Exposing the seed to a current of heated air above 40°C. Heating air reduces its relative humidity (RH). High temperatures are inevitably involved when this method is used.

ii. Placing the seed in an atmosphere which is at low relative humidity. The relative humidity of the air is reduced by a process other than that of simply heating it. Usually the temperature is maintained below 30°C.

The heated air method has dangers if a thermostat should fail and the seed overheats; however, moisture-extraction ovens can be purchased with a built-in safety thermostat, in addition to the main thermostat. This method also only deals with relatively small quantities at a time because of the oven's volume, although this disadvantage is compensated for by a fast drying time. A method which uses lower temperatures and an atmosphere which is at low-relative humidity requires more elaborate equipment and dries the seed more slowly, but it can deal with large quantities of seed (IBPGR, 1976). The use of heated air reduces seed viability more than those methods where low RH with lower maximum temperatures are used.

Thus we can divide the seed drying techniques into low temperature and high temperature methods. Generally, the former simply use heat to dry the seeds, whilst the latter methods dehumidify the air by refrigeration-reheat or by chemical methods. Nevertheless, exceptions to this rule will be seen below.

J.R. Witcombe and W. Erskine (eds.) Genetic Resources and Their Exploitation - Chickpeas, Faba beans and Lentils. ISBN-13:978-94-009-6133-3 (PB)
©1984, Martinus Nijhoff/Dr W. Junk Publishers for ICARDA and IBPGR.

Heating above 30°C

Assuming the initial moisture content of the seeds is less than 11%, the seed can be heated in a fan-ventilated oven at above 30°C until the desired moisture content is reached (in about 72 h). When the initial moisture content is higher than 11%, then two stage drying is necessary. The seed is first heated to 40°C until the moisture content has fallen below 11% and the temperature can then be raised to a maximum of 60°C.

Heating relies on two physical effects to dry the seed: the heated air has a low RH, and an increase in temperature increases the evaporation rate of the water from the seed. However, the RH of the heated air will depend on the initial conditions of the air before heating; the drier and colder the air is initially, the drier the heated air will be. Actual figures can easily be determined by using a psychrometric chart. In countries which have warm, humid climates, heating the air will not sufficiently reduce the relative humidity of the air to make this an effective method of drying seeds to low moisture contents. Consequently, although IRRI uses high temperature drying, the air needs to be dehumidified by chemical means before heating.

Equilibration with an atmosphere at low RH at low temperatures

The seeds are placed in a small room or cabinet which is moisture-vapour proof and exposed to an atmosphere maintained at about 10 to 20% RH by means of a dehumidifier. The air can be dehumidified using two main methods:

1.Chemical dehumidifiers

The best chemical dehumidifiers are those that rely on silica gel only. These dehumidifiers are absorbent types since the silica gel absorbs water from the air; the silica gel is regenerated by heating. Modern types of silica gel dehumidifiers are of the 'rotary, impregnated structure' type and have the lowest energy requirement of any type of chemical dehumidifier. Lithium chloride (LiCl) driers are also used, but LiCl requires more energy for regeneration, and is a more expensive and more corrosive chemical than silica gel.

Since a heat load is imposed on the room by the regeneration of the silica gel the room usually needs to be cooled to maintain a low temperature; however, this will depend on the ambient temperature.

Examples of such a system can be found at Braunschweig, Federal Republic of Germany and IRRI. At Braunschweig a walk-in prefabricated drying room is used. The air is dried with a LiCl drier and the air in the room is maintained at 5% RH and 20°C with cooling by an air-conditioner. A drying period of around two

weeks is used. At IRRI dehumidified air from a silica gel dehumidifier is
chilled to below 38°C before being fed into drying ovens and heated to 38°C.
The moisture content of the seeds is reduced to about 6% moisture content in 19
hours.

2. Refrigeration and reheat

In this method the air is dried by cooling and reheating. During cooling
water vapour from the air condenses and freezes on to the cooling coil of the
cooling unit. Air that has lost water in this fashion and which is reheated to
its initial temperature has a greatly reduced relative humidity. The water that
is frozen on the coil is eliminated during a defrost cycle. During this defrost
the cooling unit may be isolated from the drying room to prevent a transient
increase in RH in the room. (However, such a transient increase will only slow
down the drying of the seed). The reheat may be carried out either by using
electrical heaters or by using reject heat from the cooling unit condenser.

Reheat with electric heating

By this method, the cooling unit can easily be isolated from the room during
defrost by a mechanical damper, since the cooling unit can be external with
cooled air ducted in. Electric reheat increases running costs but the
temperature of the room can be maintained below ambient. An example of such a
system can be found at Bari, Italy.

Reheat with reject heat from the condenser
i. No provision for cooling

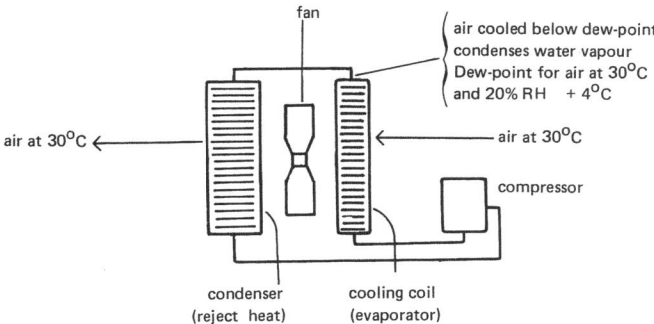

Figure 1. Schematic diagram of dehumidifier using refrigeration and condenser reject
heat.

Running costs are greatly reduced if reject heat from the condenser is used for reheat. The condenser and evaporator are located within the same room. The cooling unit is not used to change the temperature of the room. Its only effect will be to cause a slight increase in temperature from the defrost heating element and from waste heat from the compressor. The cooling unit is used to remove water from the air - as the air passes over the cooling coil it condenses (and later perhaps freezes). The air is cooled on the coil and is reheated as it passes over the condenser (Figure 1). For operation in the winter a thermostatically operated fan-heater is provided.

This system offers low running costs and the easiest maintenance. Its disadvantage is that there is no provision to run the room below ambient temperatures but, depending on the climate, this is often unnecessary. An example of the use of such a system is at ICARDA where the cooling unit is not isolated from the room during defrost, but to make such a provision is neither difficult nor expensive.

ii. Provision for cooling

Condenser reject heat may be used for reheat whilst allowing for cooling of the room. All that is necessary is that only a controlled part of the condenser reject heat is used to reheat the room. This can be done in two ways:

- The condenser is located outside the room and heat from it is ducted either into the room or outside. A thermostatically controlled damper is used to direct the heat.

- The cooling unit has two condensers attached to the compressor and evaporator. One condenser (for reheat) is located in the room, the other (for room cooling) is outside. Refrigerant is directed to either of the two condensers by means of a thermostatically controlled valve.

Design of the cold store

The design and cost aspects of cold stores have been treated at length in IBPGR (1976) and Cromarty et al. (1983), and are also considered by Hendricks and Meerman (1981). Nevertheless, many of the conclusions of these reports are open to argument, and many of the solutions adopted by gene-banks differ from those recommended.

Pre-fabricated cold stores

The purchase of pre-fabricated cold stores with a guaranteed performance is preferable to on-site construction of a cold store and the installation of cooling units. Pre-fabricated cold stores consist of 'sandwich' panels of insulating foam sheathed with metal on both sides. The panels are assembled on site with moisture proof gaskets between them to form the cold store. Such a store has a known performance in terms of heat insulation. When ambient conditions are known the required cooling capacity can be calculated to give a guaranteed performance.

It is usual to provide two cooling units for the cold store, each capable of taking 100% of the load. Air-cooled condensers, 'open' type compressors and R 12 refrigerant are all advisable for reliability and ease of maintenance.

Air-lock (ante-room)

It is important that the relative humidity of the cold store is not excessive. The more times a cold store is entered, the greater will be its RH as the warmer air that comes in with each entry is cooled. The severity of this problem varies with the temperature and humidity of the outside air. An air-lock (ante-room) to the cold store can be used to minimize the unnecessary load on the cooling units of both heat and moisture. The air in this room is cooled and/or dried. Air-conditioning is a simple and inexpensive method.

In many gene-banks this air-lock doubles as a drying room and is maintained at 20°C and around 15% RH. The drying room can also be used for seed packaging, and since personnel are working for long periods in the room it is best that it has windows and, despite the need to control temperature and moisture, it should be ventilated when the room is occupied.

Design conditions

Base collections

The preferred standard is -20°C. Since it is expensive in both capital and running costs to dehumidify air at -20°C the humidity is uncontrolled and therefore high. Note that a domestic deep freezer provides these preferred standards. Hermetically sealed containers are essential. Placing seeds at -20°C in unsealed containers would reduce their longevity because:

i.   The RH at -20°C could be nearly 100%.

ii.   The equilibrium moisture content of seeds increases with a reduction in temperature.

Active collections

Conditions of about 5°C and low seed moisture content are ideal for active collections. Although many gene-banks use sealed containers and uncontrolled RH it is advisable to dehumidify the air to 15% RH (or below).

The overwhelming advantages of controlled RH is that cheaper open containers can be used, there is no labour involved in pre-drying and packaging the seeds, and no risk of losing material due to failure of the seals on hermetic containers.

Controlling relative humidity in the cold store

Methods of controlling relative humidity were discussed above in relation to drying rooms. However, at 2° C different considerations apply. It is generally agreed that, at low temperatures and low RH, chemical dehumidifiers are to be preferred to refrigeration-reheat methods. This is because, under these conditions, refrigeration techniques require very low temperatures since the dew-point is very low. The condensed water will freeze on the cooling coils and defrosting devices will be required. The normal method is to have two cooling units so that one can run whilst the other defrosts. Since provision is made for one unit to be out of order, then at times intermittent running of one unit, with an increase in RH during defrost, will occur. The various methods are discussed below.

i.   Cooling and reheat with electric heating.

   This is probably the most expensive method in terms of running costs but it is argued, by some contractors, that it is the most reliable.

ii. Cooling and reheat using condenser reject heat.

   This method is also reliable and since running costs would be considerably less than with electric reheat it is to be preferred.

iii. Silica gel dehumidifiers.

Two examples of dehumidified cold stores are:

   - At the PBI, Cambridge, silica gel driers successfully maintain the cold store at below 10% RH despite high ambient RH (often in excess of 80% RH).

   - At IRRI, where ambient conditions are unfavourable (hot and humid), cooling and reheat using electric heaters is employed. One room is maintained at -10° C and 30% RH, and another at 4° C and 45% RH. The -10° C room is located within the 4° C room. The refrigerant plant required to dehumidify the air in the -10° C room is small, since penetration of moisture from outside is very low.

The IBPGR working group (1976) argued that using conventional refrigeration techniques, the RH of the store will automatically be controlled below 70% and can possibly be as low as 40%. This is a misleading statement since in many cold stores the RH will be 100%. The RH achieved by a cold store which is not deliberately designed to be maintained at low RH will depend on, amongst others:

i.   Moisture permeability of the cold room.

ii. Number of door openings, product load and load from people in the store.

iii. Ambient temperature and RH.

iv. Heat insulation. The poorer the heat insulation the more likely the RH is to be low, since poor insulation is equivalent to reheat.

v.   Evaporator temperature (depends on refrigerant used), coil design (i.e. its depth, fin spacing etc.), frequency of defrost cycle and whether another cooling unit is operating during the defrost cycle..

The running costs of dehumidifying a cold store

The cost of running a dehumidified cold store is touched upon in Cromarty et al. (1983). A minimum moisture extraction load of 2.5 kg/day/100 m^3 was assumed. From figures presented the maximum cost of dehumidifying a cold store is 9.4 times the cost of refrigerating a store to 5°C.

Calculations made for ICARDA in the design of a cold store for active collections indicate running costs considerably less than the maxima suggested by Cromarty et al. (1983). Unfortunately, the running costs of dehumidifying a cold store are affected by a large number of variables which, for a rotary absorbent dehumidifier, include:

i.   The quantity of water to be removed from the air per unit time to maintain design conditions.

ii.   The desired RH of the store (the lower the RH the greater the running costs).

iii. The desired temperature of the store (lower temperatures increase running costs).

iv.   The temperature and the RH of the ambient air taken in and heated to reactivate the silica gel (high temperatures and RH increase running costs).

v.   Air flow through the dehumidifier (up to a point, reducing air flow increases the moisture extraction capacity and reduces energy costs).

The quantity of water to be removed from the store is mainly determined by:

i.   The unintended ventilation of the store and the moisture content of the incoming air. Unintended ventilation is very low from a well installed pre-fabricated cold store.

ii.   Intended ventilation (usually zero).

iii. Ventilation due to door openings. The incoming air will be from the ante-room which has a controlled environment. If the air in the ante room is maintained at 20°C and 10% RH the air has a moisture content of .75 g/kg compared to .23 g/kg for air at 4°C and 10% RH.

iv.   Moisture from the seeds in the store (the product load) and from people working in the store. The load from the seed is negligible since these should be pre-dried at temperatures higher than those in a cold store. At higher temperatures the cost of removing water is lower. (In a cold store without dehumidification the seeds need to be dried before placing in sealed containers; the cost of drying the seed is therefore incurred whatever the system used).

Removing a kilogram of water from air at about 4°C and 10% RH will (in the case of an absorption rotary dehumidifier, and manufacturer's figures calculated for the worst ambient conditions for the air used to regenerate the silica gel) impose a heat load on the cold store of approximately 11.5 kW. This heat then needs to be removed from the room by the cooling units so the energy costs of removing water are considerable. The energy costs of dehumidification are therefore largely determined by the amount of water that has to be removed. The greatest source of water is from unintended ventilation so the air conditions outside the cold store are critical. Examples of calculating the moisture load are given below:

Example 1

1. Volume of store $\qquad$ 100 m^3
2. Unintended ventilation (changes per hour) $\qquad$ 0.15 ch/h
3. Outside (ambient) conditions 40°C 30% RH $\qquad$ 13.5 g/kg
4. Cold room conditions 4°C 10% RH $\qquad$ .5 g/kg

Moisture load per hour equals the volume of air changed (m^3) times the specific gravity of the air (1.2 kg/m^3) times the difference in moisture content between the outside air and the cold room air.

$$100 \times 0.15 \times 1.2 \times (13.5 - .5) = 234 \text{ g/h}$$
$$= 5.6 \text{ kg/d}$$

Example 2

1. Volume of store $\qquad$ 100 m^3
2. Unintended ventilation (changes per hour) $\qquad$ 0.015 ch/h
3. Outside (ambient) conditions 20°C 40% RH $\qquad$ 6 g/kg
4. Cold room conditions 4°C 10% RH $\qquad$ .5 g/kg

$$100 \times .015 \times 1.2 \times (6 - .5) = 9.9 \quad \text{g/h}$$
$$= .24 \text{ kg/d}$$

Reduced unintended ventilation and better ambient conditions (which can be created by air conditioning) drastically reduce the moisture load. Even without air conditioning, but with improved unintended ventilation, the load would be reduced to .56 kg/d. It can often be energy efficient to air condition the room housing the cold store, depending on its size and construction and the ambient conditions.

Since the unintended ventilation rate has such a huge effect on the moisture load and consequently the running costs it is not surprising that greatly differing estimates are given on the running costs of dehumidification. At ICARDA, estimates are that dehumidification running costs will be approximately equal to those of refrigeration. It must be remembered that these running costs are still very small in relation to the total cost of running a gene-bank facility.

Determining the RH in the cold store

Unfortunately, the RH of the cold store is difficult to monitor because a sling-psychrometer tends to be inaccurate at low temperatures, since the difference between the wet and dry bulb temperature is small. A hair-operated hygrometer has to be corrected for temperature and requires frequent readjustment and replacement of the humidity element. (Hygrometers are usually calibrated at $+20^{o}$C, at 0°C and -5°C, 8% RH and 10% RH, respectively, need to be subtracted from the reading).

A practical solution is to test the moisture content of seeds placed in open containers in the cold store, since the concern with RH is that seeds when in an improperly sealed container may equilibrate at an excessively high moisture content. In a cold store at 70% RH at 0°C (assuming the seeds were initially at a low moisture content and are therefore absorbing water) many food legumes will equilibrate at 13 or 14% moisture content which is unsatisfactorily high; whereas at 40% RH, the moisture content will 8 or 9%. Germination and moisture tests are made on the seeds kept in open containers, and are compared with seeds kept in closed containers.

A more expensive, but definitive solution, is to use a hygroceramic probe connected to an electronic chart recorder. Probes can be purchased which can record RH down to 0%.

Shelving

Shelf size, capacity and spacing

The capacity of a cold store is simply calculated once the capacity of each shelving unit is known. An example of a calculation to determine the capacity of a shelving unit intended for a 3 m high room is as follows:

- The shelves are 94 cm long by 45 cm deep (this is a standard size).
- 11 shelves will give 10 spaces 20 cm deep in a height of 240 cm (each shelf is 3 cm thick). 60 cm is allowed for air flow above the units, and 7 cm below.
- Each shelf will hold 100 containers of about 500 cm^{3} capacity (diam 9 cm, height 9 cm) arranged two high on the shelf. (It should be noted that placing containers two-high on a shelf increases the capacity by about 40%, when compared to containers one-high on each shelf. This is because less space is occupied by shelves and the gaps between containers and shelves).
- Each shelf unit of 11 shelves will therefore hold 1,000 containers (or 1,100 if the top-most shelf is used with a reduction in space for air-flow).

The cold store capacities given below assume gangways of 70-100 cm and optimum layout of units in the store. We thus have a range of capacities

depending on the size and shape of the store as follows:
- Static shelving: about 370-405 0.5 litre accessions per m^3.
- Mobile shelving: about 505-555 0.5 litre accessions per m^3.

Mobile versus static shelving

A number of incorrect statements have been written on this subject.

In IBPGR (1976) it is stated that mobile shelving could double the capacity of the cold store. This is simply untrue. The maximum space savings possible, with realistic shelf widths and gangways, are in the order of 50%. (It should be noted that the advantage of mobile shelving increases with a reduction in cost efficiency of the static system. An inefficient static system has wide gangways and narrow shelves).

The report of Hendricks and Meerman (1981) compares the cost of mobile and static shelving systems. These authors state that static shelving is cheaper than a mobile system even though this contradicts the conclusion of the IBPGR working group. Hendricks and Meerman's statement is incorrect because:

i.   They only took the cost of the pre-fabricated panels for the cold store into account. However, capital costs for refrigeration plant are less, per accession, with mobile shelving as less volume needs to be cooled per accession.

ii. In their example there was wasted space in the room containing the mobile shelves which could have been used for additional static shelving.

iii. The cost differential they used between static and mobile shelving was too high.

A cost comparison of static and mobile shelving systems

Since the cost of cold stores does not increase linearly with volume it is better to calculate the advantage / disadvantage of mobile and static shelves by considering a fixed size of room and calculating the respective cost per accession. It can be seen in the following example that the cost per accession is reduced by around 22% with mobile shelving.

1. Static shelving (see Figure 2)
   Cold store (12x7x3 m) at $475/m^3   = $120,000
   93 static units (1x.5 m) at $88   = $ 8,184
                                                 $128,184

   Cost per accession (1,000 per unit)   = $ 1.38

2. Mobile shelving (see Figure 2)

Cold store (12x7x3 m) at $475/m^3 = $120,000

127 mobile units (1x.5m) at $190   = $ 24,130

$144,130

Cost per accession (1,000 per unit)   = $ 1.13

This cost saving will always apply if mobile shelving costs about twice as much static shelving, and mobile shelving costs are about 20% of that of the cold store. These conclusions will therefore remain valid despite inflation as long as cold store and shelving costs increase at about the same rate.

The cost per accession with the mobile shelving only becomes equivalent to that of the static shelving when the cost of the cold store falls to very low levels.

'Rules of thumb' for comparison of static and mobile shelving:

With static shelving as a baseline:

i.   Changing to mobile shelving increases the capacity of a cold store by about 35-40%.

ii.   Keeping accession number constant, the area of a cold store can be reduced by about 30% if mobile shelving is adopted.

With mobile shelving as a baseline:

iii. Changing to static shelving reduces the capacity of the cold store by about 30%.

iv.   Keeping the accession number constant, about a 40% increase in cold store area is required if static shelves are adopted.

v.   The increase in the storage capacity of a store by using mobile shelving is largely independent of the size of the cold store.

vi.   Savings in cost made by using mobile shelving are not independent of the size of the cold store, since the cost per unit volume of a store decreases with an increase in the size of the store. This is why Cromarty et al. (1983) have two, apparently contradictory, conclusions that (a) with a 100 m^3 of cold room for both the static and the mobile system, the costs per accession are reduced with mobile shelving (b) costs are approximately the same if two stores of the same capacity are compared i.e. a 100 m^3 store with mobile shelving and a 181 m^3 store with static shelving. The equivalent cost is because the larger store, which houses the static system, is cheaper per unit volume than the small store. It must be noted that if a larger capacity is required then the mobile system is, again, cheaper.

Actual savings, from a financial viewpoint, will usually be in the region of 25%.

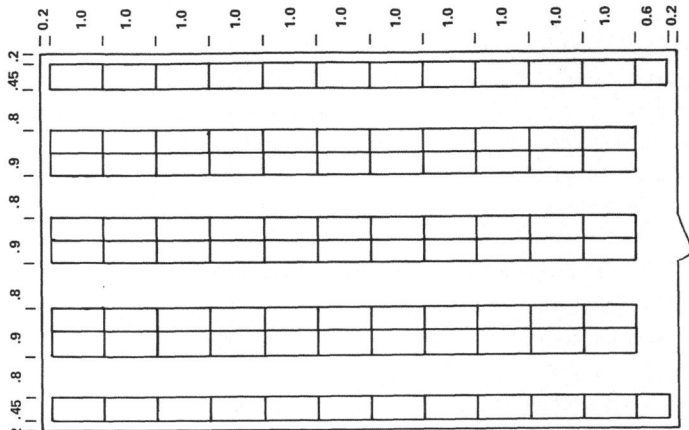

Floor plan of static system—89 units

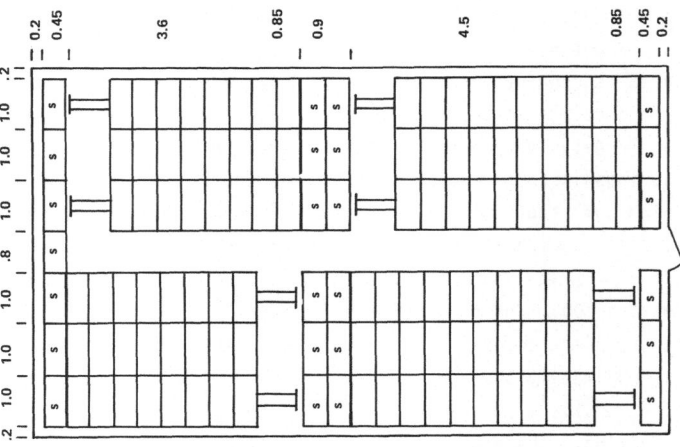

Floor plan of mobile system—129 units

Figure 2. Comparison of static and mobile shelving systems for a cold store of 12m x 7.2m
(s = static shelving units in the mobile system).

Conclusions

In general mobile shelving saves both capital and running costs. Nevertheless, in the case of active collections where running costs are less, the extra convenience of a static shelf system is probably worth the increase in cost of 25% over the mobile system. For base collections there is no doubt that the mobile system is best. Its only possible disadvantage is that mobile shelving could give problems at -20°C, but this is unlikely.

Choice of containers

The ideal container is:
- Capable of being sealed hermetically with 100% reliability.
- Impermeable to gas and water vapour.
- Transparent (so contents can be seen).
- Not fragile.
- Resealable.
- Relatively inexpensive.

The variety of containers used in different gene-banks include: Screw-top cans, permanently sealed cans, screw-top glass jars and vials, permanently sealed glass containers, laminated-foil pouches and plastic bottles. Hence it is obvious that there is no such thing as an ideal container.

Screw-top cans are commonly used and are re-sealable and unbreakable containers. However, they are opaque and, if the lid is poorly tightened, may not have a perfect seal. The IBPGR working group (1976) recommended machine-sealed cans, but although there may be provision for resealing them, they are less convenient than screw-top cans. Nevertheless, for base collections, machine-sealed containers are desirable.

Screw-top glass jars are good containers. The contents are readily seen, and seal failure can be detected by using indicator silica gel or silica gel with cobalt paper. Labelling is simplified as internal paper labels are clearly visible. However, glass jars are breakable and if several are dropped and broken, the seeds of different accessions can become inextricably mixed. It is worth considering the use of polythene bags with a rapid closing device placed inside or outside the jars to overcome this disadvantage. The cost of jars varies greatly depending on their local availability.

Glass jars should be used to store small sub-samples that are in polythene or laminated-foil pouches; this facilitates stock control, since they can be seen without unpacking them.

For small seeded species, e.g. lentil, 28 ml glass vials (Universal bottles or McCartney bottles) are ideal containers. These glass vials have an aluminium screw-top with a natural rubber liner. Tests at the Royal Botanic Gardens, Kew,

UK, have shown them to be impermeable to water vapour at $-20\,^{\circ}C$ over long periods. Labelling is done with small paper labels placed inside the vials.

Laminated-foil pouches that are heat sealed are often used. However, it is known they are slightly permeable (Freire and Mumford, 1981) and they cannot safely be resealed even though this procedure is recommended. Foil pouches require careful handling to avoid damaging them. There are varying qualities in foil pouches and those with a heavy gauge specification should be used. It is surprising that polythene (500 gauge) is not markedly inferior to laminated-foil pouches (Freire and Mumford, 1981).

Depending on the purpose and quantity of the material stored, different containers are used:

PURPOSE	POSSIBLE CONTAINERS
Active collections	Screw-top cans or jars (500 or 1000 ml capacity). In a dehumidified store 'open' containers can be used e.g. paper bags.
Base collections	Permanently sealed cans or glass containers. Screw-top glass jars with indicator (250 to 1000 ml capacity)
Small seeded species. Small base collections. Small sub-samples for sowing or distribution to active collections	Screw-top glass vials (28 ml capacity). Laminated-foil pouches. Polythene bags (stored in cold store inside glass jars). Laminated-foil pouches.
Distribution of sub-samples for long term storage (base collections).	Heat-sealed laminated-foil pouches (preferably they are repacked or placed inside other containers by recipients).

Arrangement of containers in the cold store

Containers can be stored directly on the shelves, but this is an inconvenient method of handling them. Instead, cardboard or wooden trays (i.e. boxes without lids) are used which hold the containers in a single layer, e.g. ten 500 ml containers per tray. The trays are stored two high on the shelves (only one high markedly reduces the capacity of the store, more than two high is less convenient) and access to any particular accession is gained by moving a maximum of one tray. To rapidly find any accession in the cold store the accessions are labelled:

- On the container with the accession number, the species and the year of harvest.
- On the lid with the accession number only.
- On the tray, so that it is clearly visible when stored in the cold store, with accession number, species and year of harvest.

This system is also designed to be convenient when material is removed to the laboratory. When the trays are laid in order on the bench, all the accessions are readily found by means of the numbers on the lids of the containers, which are in a single layer on each tray.

Each position in the cold store (most conveniently the position of a tray) is allocated a number, and this is written both on a plan and on the shelf in the cold store. No matter how many cold stores or buildings, a single number is sufficient if each store is initially allocated a large enough set of numbers and the numbers follow consecutively from store to store.

It may well be found that the location numbers are unnecessary if the containers are stored in some degree of order in the cold store. The containers can be arranged in the following manner:

e.g. Shelf 1: Lentil 1986, Lentil 1987, Lentil 1988.
     Shelf 2: Faba 1986, Faba 1987, Faba 1988.

Within each 'year block', e.g. Lentil 1986, the accessions are arranged in order. Thus, the containers of the same accession which has been grown in different years are not found together, but in their appropriate blocks according to their harvest years. Within a 'year block' all types of seed are entered - multiplied seed, collected seed and seed received from donors. The advantage of this system is that all seeds acquired during the current year can be stored without moving those of previous years.

The task of maintaining the containers in order is made easier by allocating a shelf, or group of shelves, to a crop category. Containers can then be added to the existing stock in subsequent years without moving material to make space available. If after a number of years, the allocated space for a particular crop is exhausted, a fresh block can be started in another part of the store. This avoids the necessity of rearranging stocks, and does not create a system that is so complex that location numbers are required.

References

Cromarty, A.S., Ellis, R.H. and Roberts, E.H. 1983. The Design of Seed Storage Facilities for Genetic Conservation. AGPG:IBPGR/82/23, Rome: IBPGR.

Freire, S.M. and Mumford, P.M. 1981. Packaging seeds for storage - a preliminary report. Plant Genetic Resources Newsletter 48: 13-17.

Hendricks, P. and Meerman, H.J. 1981. Report of a Survey of Gene Bank Facilities and Preparation of Seed for Storage in Europe. Technische en Fyscsiche Dienst Voar de Handbouw, Wageningen.

IBPGR. 1976. Report of IBPGR Working Group on Engineering, Design and Cost Aspects of Long Term Seed Storage Facilities., AGPE:IBPGR 76/25, Rome.

4. DOCUMENTATION OF GERMPLASM COLLECTIONS BY COMPUTER

J.R. Witcombe

FAO/IBPGR, South West Asia Program

W. Erskine

Food Legume Improvement Program, ICARDA, Aleppo, Syria

Introduction

The IBPGR has made considerable progress in establishing standard lists for the description of germplasm accessions. Such lists are referred to elsewhere. See, for example, the IBPGR collecting form (Appendix 2, Chapter 1), and draft descriptor lists for faba beans (Appendix 10, Chapter 13) and lentils (Appendix 11, Chapter 17).

It is not, therefore, our aim to deal with the standardisation of the description of germplasm material. Instead, the computerisation at ICARDA of two large data sets, for chickpea and lentil germplasm, is described from a practical viewpoint. Points of interest that are independent of a particular computer or computer program are discussed.

The computerisation of a data set and its subsequent use follows many steps, and some of the main ones for both passport and evaluation data are listed, in sequence, below:

1. Initial data verification and reorganisation
2. Data entry into the computer
3. Data verification
4. Editing
5. Re-verification
6. Report writing
7. Analysis
8. Querying the data base

Passport Data

The poor quality of the passport data provided by most donors presents great difficulties. Hence, the initial phase of reorganising the donors passport information is the task that occupies the greatest time.

The types of problems encountered can be summarised as:

i.   A lack of information

ii.   The task of assigning the information provided to the correct descriptor.

The form used at ICARDA in accession books is shown in Appendix 3, Chapter

J.R. Witcombe and W. Erskine (eds.) Genetic Resources and Their Exploitation - Chickpeas, Faba beans and Lentils. ISBN-13:978-94-009-6133-3 (PB)
©1984, Martinus Nijhoff/Dr W. Junk Publishers for ICARDA and IBPGR.

1. This differs in several ways from the IBPGR passport information descriptor list.

When information is lacking the only recourse is to write to the donor requesting further information. It can be seen from Table 2 that, despite this approach, in many cases vital descriptors are not known.

Table 2. Proportion of missing data in the lentil passport information.

Descriptor	No of known values *	Percentage unknown
Collecting Organisation	3962	26
Collector	1136	79
Collector's number	4345	20
Collection date	204	96
Country of origin	5372	10
Town/Province	1359	75
Latitude	45	99
Longitude	45	99
Altitude	45	99

* From a total of 5424 accessions

The allocation of information to the correct descriptor is often difficult. Decisions have to be made in the absence of complete information, as to whether the donor's number is also the collector's number. In turn, this decision will also often affect the recorded country of origin of the accession. Since the collector's number should be a unique identifier accompanying the accession wherever it is sent, its correct identification is important. Only too frequently, when an accession is passed from organisation to organisation, it acquires new numbers ('synonyms' - the accession numbers assigned to it by different organisations) and the original collection number is lost.

Numerous other confusions are possible, e.g. is a name a cultivar name or the place of collection? Is a name a town or a province or an indication, in a foreign language, of some of its characters, e.g. large seed size or quick cooking. Misspellings also occur. For example, 'spots Alpinse' was corrected to 'Spate Alpinse' which means late highland lentil; this is a collector's name, which is also descriptive of its maturity and origin, and is not, as might be expected, a place name or variety name.

Once the data have been satisfactorily allocated to the descriptors, they can be entered on the computer. Ideally, this is done with a data entry program that checks that each descriptor is within the limits that have been specified, e.g. that the date of collection is a four digit number within the limits of perhaps, 0130 and 1284 (January, 1930 to December 1984).

When the data have been entered they are checked according to the nature of the descriptor (either alphanumeric, or integer, or real number).

i. Alphanumerics

An example of such a descriptor would be 'town and province'. The data are sorted in alphanumeric order. In this way misspellings (or various legitimate spellings of the same name) can be identified, e.g. discrepancies such as Isphahan, Isfahan are immediately found. Erbil and Irbil are less easily found, since they occur in a different part of the list; moreover, they require checking to see if they are two different legitimate names, or misspellings or variants of the same name. The index of a standard atlas is invaluable in checking the spelling of locations. It is easy to visualise how time consuming this procedure can be, even though the computer is used to find 'Erbil' and 'Irbil' in the passport data file so that the relevant accession numbers can be identified.

ii. Integers

These can be sorted to find the number of distinct items found and their values and frequencies. Such a sort is useful in identifying implausible outliers (i.e. values that are much larger or smaller than appear reasonable) but a good data entry program will do this at the time of data entry (each item is checked against the minimum and maximum expected values for that descriptor).

iii. Real numbers

A statistical summary which includes the minimum and maximum can be produced; this again serves the purpose of identifying implausible outliers.

It must be noted that the above check for integer and real numbers will only determine that the data set, after checking and editing, will consist of legitimate values. However, many accessions could still have false values assigned to them. There are two ways of checking this.

i.   The data can be entered twice, and the two files produced then compared on the computer. This is a feasible method when fast and expert data entry staff are available, and computer time is unrestricted.

ii. The data can be checked using the time-consuming and time-honoured method of comparing, by manual means, the original data to a hard-copy output of the computer file. This procedure has been followed by ICARDA.

Once errors have been identified, the computer files are edited.

Unfortunately, if there are many corrections, then the data need to be rechecked to discover possible errors introduced during the editing process. In fact, the procedure becomes iterative.

Listing data

At this point a listing of the files can be produced since a computerised, checked version of the accession book is now available. Example pages of chickpea and lentil passport and evaluation data (Singh et al. 1983, Erskine and Witcombe, 1984, respectively) are shown in Table 3, and Tables 9 and 10, Chapter 10.

Listing the files is done by using a 'report generator'. The report has to be presented in columns and has to fit the width of the page, which is usually resticted to a maximum of 132 characters. Although the computer can store a descriptor of any length, it is at this stage that long descriptors will not fit easily into the format of a report. There are several solutions:

i.   Abbreviations. When an alphanumeric sort on a descriptor has been done to check the data, the largest descriptors are readily identified by scanning the list. These can be abbreviated, if this is possible without losing the meaning of the descriptor.

ii. Using more than one line per accession. The data of each accession may occupy several lines. The report can be presented with the lines for each accession following consecutively in the report. This method destroys the clear columnar presentation. Alternatively, the data are separated so that the accessions are listed, one line per accession, for more than one set of descriptors, e.g. the chickpea passport data was listed as 'donor and origin' and as 'synonyms' (Tables 9 and 10). In this method, all the data for any one accession are not found together, but in separate listings or reports.

Even if the data for the accessions are spread over more than one line then abbreviations will still probably have to be used.

iii. Cross-referencing. In the case of the lentil collection, few accessions had detailed collection site data, or other information beyond the nine standard descriptors used (Table 2). Those accessions with such information had a 'yes' entered in a column labelled 'index'. A separate report for the additional descriptors was produced for the accessions marked 'yes' .

Analysis of passport data

The passport data base can now be exploited in many ways. For example, frequency tables of particular descriptors can be produced. One of the most important descriptors, in this respect, is country of origin which is shown for

Table 3. Sample pages of passport information from the ICARDA lentil collection.

The following abbreviations are used in the headings:

ILL   International Legume Lentil/the accession number
Don Org   Donor organisation
Don No   Donor number
Col Org   Collecting organisation
Coll No   Collection number
CDat   Collection date
Cou   Country of origin
RDat   Date received by ICARDA/ALAD
Ind   Index
Lat   Latitude
Long   Longitude
Alt   Altitude (m)

The index column indicates if there is additional information for that accession, i.e. if there are any remarks, or if the latitude, longitude and altitude are recorded. These latter descriptors are printed separately, so that the first ten descriptors can be printed with one line per accession.

ILL	Don Org	Don No	Col Org	Collector	Coll No (Pedigree)	CDat	Cou	Province or Town	RDat	Ind
4726	INIACOR	LE 103	INIACOR	J.I.CUBERO	LE 103	0079	ESP	JAEN	1180	*
4727	INIACOR	LE 104	INIACOR	J.I.CUBERO	LE 104	0079	ESP	MALAGA	1180	*
4728	INIACOR	LE 105	INIACOR	J.I.CUBERO	LE 105	0079	ESP	GRANADA	1180	*
4729	INIACOR	LE 106	INIACOR	J.I.CUBERO	LE 106	0079	ESP	GRANADA	1180	*
4730	INIACOR	LE 107	INIACOR	J.I.CUBERO	LE 107	0079	ESP	CORDOBA	1180	*
4731	CDC	PI 179313	CDC			-	TUR		1180	-
4732	CDC	PI 345631				-	SUN		1180	-
4733	CDC	PI 368650				-	YUG		1180	-
4734	CDC	PENZISKAJA 14			PENZISKAJA 14	-	SUN		1180	-
4735	CDC	G 103	CDC	A.SLINKARD	G 103	-	CAN		1180	-
4736	CDC	G 118	CDC	A.SLINKARD	G 118	-	-	-	1180	-
4737	CDC	TE BLK F12 406M	CDC	A.SLINKARD	TE BLK F12 406M	-	-	-	1180	-
4738	CDC	ESTON	CDC	A.SLINKARD	ESTON	-	USA	-	1180	-
4739	CDC	COMMERC. CHILEAN	CDC	A.SLINKARD	COMMERC. CHILEAN	-	USA	FALOUSE	1180	-
4740	COOP	ANICIA	COOP	R.DELORE	ANICIA	-	FRA	LOIR ET CHER	1180	-
4741	IBPGR	LENS 9	IBPGR	R.P.CROSTON	LENS 9	0580	YEM	MAWER	1080	*
4742	IBPGR	LENS 15	IBPGR	R.P.CROSTON	LENS 15	0580	YEM	AL SHARAF	1080	*
4743	IBPGR	LENS 22	IBPGR	R.P.CROSTON	LENS 22	0580	YEM	AL HAJAR	1080	*
4744	IBPGR	LENS 27	IBPGR	R.P.CROSTON	LENS 27	0580	YEM	AL KHABBAS	1080	*
4745	IBPGR	LENS 29	IBPGR	R.P.CROSTON	LENS 29	0580	YEM	AGABA	1080	*
4746	IBPGR	LENS 37	IBPGR	R.P.CROSTON	LENS 37	0580	YEM	AL HASSAN	1080	*
4747	IBPGR	LENS 42	IBPGR	R.P.CROSTON	LENS 42	0580	YEM	BAIT AL SARAIMI	1080	*
4748	IBPGR	LENS 48	IBPGR	R.P.CROSTON	LENS 48	0580	YEM	JUBAIR	1080	*
4749	IBPGR	LENS 78	IBPGR	R.P.CROSTON	LENS 78	0580	YEM	SANAA	1080	*
4750	IBPGR	LENS 195	IBPGR	R.P.CROSTON	LENS 195	0580	YEM	ZUMRAIN SABIR MAWADEM	1080	*
4751	IBPGR	LENS 261	IBPGR	R.P.CROSTON	LENS 261	0580	YEM	IBB\YAREM	1080	*
4752	IBPGR	LENS 299	IBPGR	R.P.CROSTON	LENS 299	0580	YEM	JABAL AL KHADRA	1080	*
4753	IBPGR	LENS 309	IBPGR	R.P.CROSTON	LENS 309	0580	YEM	AL HOUSAIN	1080	*
4754	IBPGR	LENS 324	IBPGR	R.P.CROSTON	LENS 324	0580	YEM	IMAN	1080	*
4755	IBPGR	LENS 365	IBPGR	R.P.CROSTON	LENS 365	0580	YEM	AL KHAN	1080	*
4756	IBPGR	LENS 367	IBPGR	R.P.CROSTON	LENS 367	0580	YEM	AL KHAW	1080	*
4757	IBPGR	LENS 375	IBPGR	R.P.CROSTON	LENS 375	0580	YEM	YARIM	1080	*
4758	IBPGR	LENS 383	IBPGR	R.P.CROSTON	LENS 383	0580	YEM	KAMAN	1080	*
4759	IBPGR	LENS 390	IBPGR	R.P.CROSTON	LENS 390	0580	YEM	DHUMARA	1080	*
4760	IBPGR	LENS 394	IBPGR	R.P.CROSTON	LENS 394	0580	YEM	QAHDHAHAR	1080	*

Ind	ILL	Remarks	Lat	Long	Alt
	*	4536 HIMOHANADY	–	–	–
	*	4537 HIMOHANADY	–	–	–
	*	4538 HIMOHANADY	–	–	–
	*	4539 MA'ASHOK	–	–	–
	*	4540 MA'ASHOK	–	–	–
	*	4541 GERBETADNAN	–	–	–
	*	4542 MA'MSHOOK	–	–	–
	*	4543 AL KAYSARIA	–	–	–
	*	4544 AL GESSER	–	–	–
	*	4545 NABISADI	–	–	–
	*	4546 NABISADI	–	–	–
	*	4547 DERZIWAR	–	–	–
	*	4548 DERZIWAR	–	–	–
	*	4549 DERZIWAR	–	–	–
	*	4551 LA LIGUA ACONCAGUA	–	–	–
	*	4555 MULCHEN BIO-BIO	–	–	–
	*	4556 MATANZAS SANTIAGO	–	–	–
	*	4557 ZAPALLAR ACONCAGUA	–	–	–
	*	4558 LONGOTOMH ACONCAGUA	–	–	–
	*	4563 SAN JUAN NAVIDAD	–	–	–
	*	4564 NAVIDAD SANTIAGO	–	–	–
	*	4565 CHILEAN	–	–	–
	*	4586 QUINTERO VALPARAISO	–	–	–
	*	4591 PUPUYA TALCA	–	–	–
	*	4592 PALQUIBUDIS CURICO	–	–	–
	*	4602 SANTA FE	2841N	8139E	2140M
	*	4610 FAR WESTERN REGION	2850N	8143E	1410M
	*	4611 FAR WESTERN REGION	–	–	–
	*	4716 VENTAS DE HUELMA	–	–	–
	*	4717 PADUL	–	–	–
	*	4718 LAMALA	–	–	–
	*	4720 ALHAMA DE GRANATA	–	–	–
	*	4721 AGUILAR DE LE FRONTERA	–	–	–
	*	4722 CUEVAS BAJOS	–	–	–
	*	4723 VILLANUEVA DEL TRABUCO	–	–	–

Ind	ILL	Remarks	Lat	Long	Alt
	*	4724 EL BURGO	–	–	–
	*	4725 CUEVAS DE ST. MARCOS	–	–	–
	*	4726 VENTAS DEL CARRIZAL	–	–	–
	*	4727 ESTACION DE SALINAS	–	–	–
	*	4728 RIO FRIO	–	–	–
	*	4729 LOJA	–	–	–
	*	4730 BENAMEJI	–	–	–
	*	4741 STORED SAMPLE	1415N	4444E	2000M
	*	4742 STORED SAMPLE	1414N	4443E	2500M
	*	4743 STORED SAMPLE	1406N	4440E	2630M
	*	4744 STORED SAMPLE	1418N	4450E	2040M
	*	4745 STORED SAMPLE	1415N	4447E	2560M
	*	4746 STORED SAMPLE	1414N	4446E	2530M
	*	4747 STORED SAMPLE	1410N	4446E	2500M
	*	4748 STORED SAMPLE	1411N	4447E	2700M
	*	4749 MARKET SAMPLE	–	–	–
	*	4750 STORED SAMPLE	1332N	4400E	2300M
	*	4751 MARKET SAMPLE	1400N	4400E	–
	*	4752 STORED SAMPLE	1406N	4406E	2500M
	*	4753 STORED SAMPLE	1406N	4409E	1700M
	*	4754 STORED SAMPLE	1415N	4415E	2150M
	*	4755 STORED SAMPLE	1417N	4428E	2200M
	*	4756 STORED SAMPLE	1417N	4438E	2200M
	*	4757 MARKET SAMPLE	1418N	4423E	–
	*	4758 STORED SAMPLE	1417N	4425E	2350M
	*	4759 THRESHING FLOOR	1425N	4427E	2200M
	*	4760 STORED SAMPLE	1430N	4424E	2350M
	*	4761 THRESHING FLOOR	1438N	4419E	2200M
	*	4762 MARKET SAMPLE	1440N	4409E	2000M
	*	4763 MARKET SAMPLE	1400N	4400E	2000M
	*	4764 STORED SAMPLE	1527N	4408E	2400M
	*	4765 STORED SAMPLE	1527N	4408E	2400M
	*	4766 STORED SAMPLE	1524N	4407E	2400M
	*	4767 STORED SAMPLE	1510N	4400E	2800M
	*	4768 STORED SAMPLE	1510N	4400E	2800M

chickpea in Table 8, Chapter 10, and for lentil in Table 27, Chapter 17. These frequency tables enable gaps to be identified in the collection where countries are under-represented.

The passport information from a particular country may also be required in planning for future collecting trips (Witcombe, Chapter 1). With the data on location of collection available, maps of previous collections can be made, in order to avoid a duplication of collecting effort.

It is of interest that, because chickpea and lentil are under-exploited legumes having received little attention from plant breeders, their collections have been shuttled around the world to different gene banks less than for other more 'important' crops. As a consequence the countries of origin of their accessions are relatively easier to ascertain.

The information on donor organisation is most useful, not least because it identifies those organisations which have many accessions in common with the collection. In the case of loss of seed of accessions the most likely source of replacement seed is known. The Egyptian national lentil collection was added to the international lentil collection in 1972. Since that date some accessions have been lost by the national program. A listing of accessions in the international lentil collection originating from Egypt helps identify those accessions to be sent back to Egypt to replenish the national collection.

The degree of duplication in the collection can be identified by performing an alphanumeric sort on the collection (pedigree) number and the donor number. In lentils about 10% of the accessions are duplicates. However, from experience, such 'paper' duplicates are not genetically identical. For example, the lentil cultivar Penjinskaya from the USSR is represented four times in the ICARDA Lentil collection, but each sample has different seed. This can be because of past mis-labelling or of different histories of selection. Nevertheless, a comparison of the evaluation data of duplicates is valuable. Those accessions between which there is little difference for highly heritable characters could be pooled, or one of them could be discarded from the active collection and maintained only in the base collection.

A search for the collection or pedigree number of newly acquired germplasm material can prevent, at source, the entry of duplicates into the collection.

Evaluation data

The evaluation data are the starting point for effective germplasm utilization.

Individual descriptors are analysed by providing a statistical summary (Table 4, Figure 3) and by listing selected, extreme accessions (Table 5).

At ICARDA, a further descriptor, country of origin as a numeric code, has

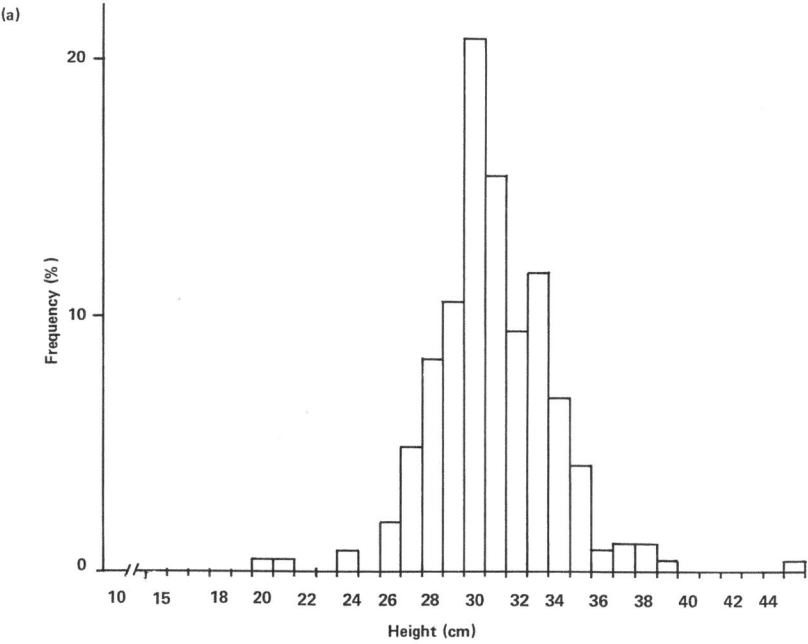

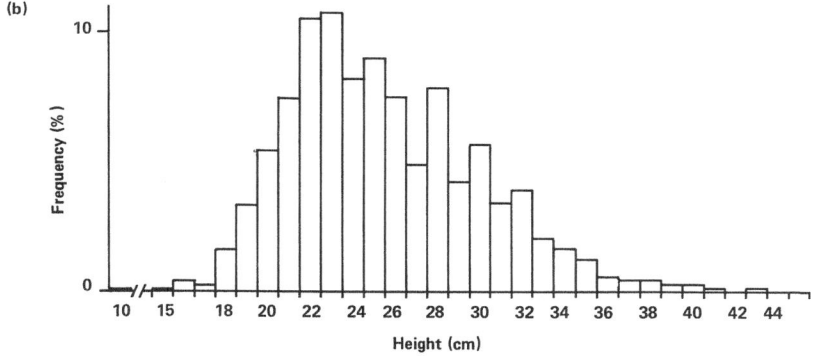

Figure 3. The distribution of plant height (cm) in the lentil germplasm collection for (a) large seeded accessions (≥ 4.5g/100 seeds) and (b) small seeded accessions (< 4.5g/100 seeds).

Table 4. Summary of statistics on plant height in the lentil
collection.

Statistics	All accessions	Large seeded	Small seeded
Mean	26.5	30.9	25.4
Mode	23.0	30.0	23.0
Minimum	10	20	10
Maximum	45	45	43
Range	35	25	33
Variance	21.62	7.88	19.22
Coeff. of var.	17.6	9.1	17.2
Kurtosis	-0.36	3.17	0.17
Skewness	0.26	0.36	0.55
Number of obs.	1370	263	1107

Table 5. Accessions in the lentil collection with a height
greater than or equal to 36 cm.

Accession numbers				
ILL 36	ILL 37	ILL 92	ILL 103	ILL 175
ILL 177	ILL 201	ILL 292	ILL 446	ILL 447
ILL 448	ILL 793	ILL 794	ILL 795	ILL 796
ILL 799	ILL 809	ILL 810	ILL 811	ILL 812
ILL 813	ILL 815	ILL 816	ILL 922	ILL 924
ILL 926	ILL 999	ILL 1141	ILL 1476	ILL 1751
ILL 2313	ILL 4349			

been entered on the evaluation data base. This enables an analysis of variance by country of origin to be carried out (Table 13), or various multivariate analyses to be done, such as discriminant analysis by country of origin, or principal component analysis.

Analyses of the variation by country of origin show the geographic distribution of the variation of the crop. This can be useful in studies of crop evolution, and in the exploitation of the germplasm. For example, lentils collected in Afghanistan, and now in the ILL collection, are particularly late to flower and late to mature. As a consequence, future germplasm introductions for Afghanistan from the international collection should be of a comparable late maturity to be adapted there. This approach can lead to a more rational utilization of existing collections.

The power of the computerised data base lies its ability to rapidly answer complex queries for particular genotypes, e.g. early maturing, frost-resistant, tall genotypes with tall, lowermost pods. Correlations between descriptors in the evaluation data are useful to those wishing to query the data base. A correlation indicates the probability of finding particular values of two descriptors. Thus, if 100-seed weight is negatively correlated to the number of seeds per pod, there is little likelihood of finding large seeded accessions with a high number of seeds per pod.

Conclusions

There is no doubt that to produce an accurate, carefully-checked data base for a germplasm collection is a time-consuming task. However, once this is done gaps in geographical representation can be identified in a collection, accessions can be categorised and selected, and the usefulness of the collection is greatly increased.

The major work of computerisation is the initial cleaning of the data and checking it after it has been entered. Although computer programs can reduce the initial number of mistakes which occur during data entry, by not accepting unreasonably high and low values, there is no quick and easy method of cleaning the data in the first place.

Finally, a word about the hardware and software used to computerise genetic resources data. For the large data sets of the chickpea and lentil collections at ICARDA it was found that a stand-alone 'micro-computer' (a North Star Horizon with 64K of memory and a disk drive with a capacity of 10 megabytes) was inadequate for the task, so the data were transferred to a larger computer. The main limitation of the micro-computer was not memory size or storage capacity for the data, but its speed in processing large sets of data. However, a further limitation was that complex analyses are prevented by the limited capacity of the memory to store and execute a large program.   Improvements in hardware should eventually ensure that a low-cost computer is equal to the task of rapidly handling the documentation of relatively large germplasm collections.

Whatever software is used to handle the germplasm data, it is advisable that the files can be converted for use with a standard statistical package.   At ICARDA the Statistical Package for the Social Sciences (SPSS) (Nie et al., 1975) was used to produce frequency distributions and to do various analyses such as correlations, analysis of variance and discriminant analysis.   Other centers have adopted this approach. For example, IRRI uses both SAS, and SPSS to analyse genetic resources data (Gomez et al., 1977)

References

Erskine, W. and Witcombe, J.R. 1984.   Lentil Germplasm Catalog. (in press). Aleppo, Syria: ICARDA.

Gomez, K.A., Tuazon, D. and Francisco, C.J.   Computer-based data management system for crop improvement programs.   Pages 709-725 in Proceedings of the International Conference on Computer Applications in Developing Countries, Bangkok, Thailand. Vol. 1. Jordan, T.A. and Malaivongs, Kanchit.(Eds).

Nie, Norman H., Hadlai Hull, C., Jenkins, Jean G., Steinbrenner, Karin, and Bent, Dale H. 1975. SPSS Statistical Package for the Social Sciences.   2nd Ed: McGraw Hill.

Singh, K.B., Malhotra, R.S. and Witcombe, J.R. 1983.   Kabuli Chickpea Germplasm Catalog. Aleppo, Syria: ICARDA.

5. COLLECTION, ISOLATION AND MAINTENANCE OF FOOD LEGUME RHIZOBIA

Rafiqul Islam
ICARDA, Aleppo, Syria

Introduction

Food legumes play a major role in agriculture, because they are capable of fixing atmospheric nitrogen through their association with rhizobia bacteria. Consequently they can be grown with minimum inputs in soils of low fertility. Research on symbiotic nitrogen fixation to improve legume production needs a ready supply of species and strains of Rhizobium for food legume species and the collection, isolation and maintenance of strains of these Rhizobium form an essential base to any research program.

Collection and isolation of rhizobia

Strains of rhizobia must be collected from the soil or from plant nodules growing in situ, and then isolated. Isolation of rhizobia from the soil is particularly difficult and requires considerable skill. However, the presence of rhizobia in the soil may be demonstrated simply by sowing surface-sterilised seeds (Appendix 5) directly into the soil, or by adding, under bacteriologically controlled conditions, a soil suspension to seedlings raised from surface-sterilised seeds. Nodule formation in the plants reveal the presence of rhizobia in the soil and these nodules may then be used for isolating strains. The isolation of the rhizobia from nodules is very simple (Appendix 6) and may be carried out with a minimum of facilities. When isolating, care should be taken to choose healthy pink nodules from the main root of the host plant.

The isolate, or culture, should be grown on a medium of yeast mannitol agar (YMA) in the presence of Congo Red (Appendix 7). Other bacteria tend to absorb the dye very quickly and produce coloured colonies. Most Rhizobium, on the other hand, do not absorb the dye and will produce colourless colonies (Vincent, 1981). Growth on this type of media may also be used to characterise rhizobia on the basis of their colony characteristics (size, shape and nature) or growth rate. Fast growing strains (i.e. rhizobia for lentil, faba bean and chickpea) will produce colonies within 2-3 days, whereas slow growing ones (i.e. rhizobia for groundnut and soybean) may take between 7-12 days to produce visible colonies. Once the colonies appear, repeated subculturing should be done from one individual colony to obtain a pure culture. Further tests should be made to confirm its identity as a rhizobia culture. The following tests are important and should be carried out routinely:

J.R. Witcombe and W. Erskine (eds.) Genetic Resources and Their Exploitation - Chickpeas, Faba beans and Lentils. ISBN-13:978-94-009-6133-3(PB)
©1984, Martinus Nijhoff/Dr W. Junk Publishers for ICARDA and IBPGR.

Glucose-peptone test

This test is very useful to distinguish rhizobia from other soil contaminant bacteria, especially Agrobacterium. Bromo-cresol purple indicator (Appendix 7) is used to determine the change in pH of the glucose-peptone medium on which the isolate is grown. Rhizobium may be distinguished, as they grow poorly in the glucose-peptone medium and cause little pH change in contrast to most other bacteria.

Microscopic examination

The suspected rhizobia can be examined under a microscope; the bacteria can be fixed on a glass slide by using an appropriate stain. The gram stain is most commonly used for microscopic examination of rhizobia bacteria. The procedure for gram staining is detailed by Doetsch (1981).

Authentication of the rhizobia culture

No fresh isolate should be added to a rhizobia collection until it has been fully authenticated by proper plant infection testing using the specific host legume crop. Several techniques are used for plant infection testing (Vincent, 1970). The seedlings obtained from surface-sterilised seeds are grown asceptically in nitrogen-free medium (Appendix 7), and inoculated with the suspersion of the test culture. The plants are grown for 6-7 weeks under optimum temperature and light conditions. The plants are then harvested and the roots examined for nodulation. The nitrogen fixing ability of the nodules can be measured with the acetylene reduction technique (Hardy et al., 1973). The authenticated culture should then be evaluated in the field for their nodule forming and nitrogen fixing ability.

Culturing Rhizobium spp.

Rhizobium spp. are heterotrophic and able to utilize a wide range of carbohydrates. They are aerobic and generally grow best between 25 and 30 $^{\circ}$C and between pH 6 and 7.

While the culturing of rhizobia bacteria is not difficult, considerable care must be taken to ensure that cultures are pure and uncontaminated.

It is essential that all glassware and media used in culturing operations be effectively sterilised prior to use. Contamination by other bacteria and fungi will considerably affect both the identification and use of cultures. Glassware may be sterilised by heating to 100 $^{\circ}$C in an ordinary oven for 24 hours. However, autoclaving for 10-25 minutes at a pressure of 8 atmospheres is undoubtedly preferable, as the destruction of dormant spores often necessitates rather drastic treatment.

All containers should be covered prior to sterilisation (plugged with cotton wool, or capped with metal or glass lids) to prevent contamination by particulate matter. When the containers are opened to permit culturing the operation should be performed rapidly and in as sterile conditions as possible, preferably inside a sterile cabinet or room. In this way the risk of entry of contaminating particles from the air is minimised.

Cultures should be transferred to the sterilised media using a heat resistant wire loop or needle which has been heated to red heat and cooled immediately before use. Sterilised pipettes should be used for transferring liquid cultures.

Routine complex media

Rhizobia do not grow well on the peptone media routinely used for culturing most bacteria. However, they can be particularly well cultured on various complex media made from extracts of plants or unicellular organisms. Yeast is the most generally suitable source, either as a freshly prepared extract or a recommended powder. Mannitol is often used as a carbon source but may be replaced by one of several sugars (e.g. glucose or sucrose). The compositions of yeast mannitol and other commonly used media are given in Appendix 7.

Record keeping

It is essential that the original history of each culture is accurately kept. In this way researchers can always be certain of the material they are using. A multitude of different methods of record keeping are used. Whichever method is chosen, however, the following data must be recorded:
- accession number
- date of collection
- host species
- rhizobia species
- location of collection
- soil type (including pH)
- plants nodulated:
   - i.   in tube culture
   - ii. in Leonard Jar assembly
   - iii. effectivity
   - iv. field test records

Maintaining rhizobia cultures

Once a pure and identified culture of rhizobia is obtained, it must be maintained in a stable form so that it may be used repeatedly as required. There are several ways in which rhizobia cultures may be effectively maintained.

Agar Culture

Cultures may be stored fresh in yeast mannitol agar (YMA) for at least six months if maintained at a room temperature of $15^{\circ}C$. The duration of storage may be increased by lowering the temperature, and a normal refrigerator is usually adequate for most purposes. Care should be taken to prevent the agar drying out under refrigeration and screw-capped test-tubes (McCartney or Universal bottles) are ideal in this respect.

Dried culture

The use of freeze-dried cultures for long-term rhizobia storage is undoubtedly the best method of bulk storage. Although the drying procedure is simple, it requires specialised equipment, such as a vacuum pump. Freeze-dried cultures may be stored for periods exceeding two years under suitable temperatures.

Porcelain bead method

This simple method, which does not require specialised equipment, has been introduced only relatively recently. Unglazed porcelain beads are suspended in a heavily populated culture of yeast mannitol broth and then transferred to small glass tubes. Sterilised cotton wool is added and the upper part of the tubes filled with dry silica gel. They may then be stored for future use. The longevity of rhizobia stored in porcelain beads is not yet known.

References

Doetsch, R.N. 1981. Determinative methods of light microscopy. Pages 21-33 in Manual of Methods for General Bacteriology, eds. P. Gerhardt, R.G.E. Murray, R.N. Costilow, E.N. Nester, W.A. Wood, N.R. Krieg, and G. Briggs Phillips, Washington: American Society for Microbiology.

Hardy, R.W.F, Burns, R.C. and Holsten, R.D. 1973. Application of the acetylene-ethylene assay for measurement of nitrogen fixation. Soil Biology Biochemistry 5: 47-81.

Vincent, J.M. 1970. A manual for the practical study of the root nodule bacteria. IBP 15. Oxford: Blackwell.

Vincent, J.M. 1981. The genus Rhizobium. Pages 818-841 in The Prokaryotes. Volume 1, eds. M.P. Starr, H. Stolp, H.G. Truper, A. Balows and H.G. Schlegel, New York: Springer-Verlag.

Appendix 5. Procedure for sterilising seeds.

Materials
- beakers or conical flasks
- seeds to be sterilised
- acidified mercuric chloride (1 ml conc. HCl/litre 0.1 $HgCl_2$ solution)
- 70% ethanol
- sterilised water

Procedure
- place the seeds in a suitable beaker or conical flask
- rinse the seeds with ethanol for 30-45 seconds and drain off excess
- add acidified mercuric chloride so that the seeds are covered
- shake the flask for two to three minutes (depending upon the structure of the seed coat - rough coats require longer than smooth ones) and drain off excess $HgCl_2$
- wash the seeds thoroughly (at least six times) with sterilised water

Appendix 6.   Isolation of <u>Rhizobium</u> from fresh legume nodules.

Materials

- a well nodulated legume root
- sterilised beakers or petri-dishes
- 70% ethanol
- acidified mercuric chloride (Appendix 5)
- sterilised water
- glass rod
- inoculation loop
- YMA Congo Red plate (Appendix 7)
- incubator (20-28°C)

Procedure

- select a few healthy primary nodules which appear pink or brown in colour
- wash the nodules thoroughly in a suitable container and drain off, then rinse once with 70% ethanol and drain off excess
- immerse in acidified HgCl$_2$ and drain off excess
- wash thoroughly (at least six times) with sterilised water
- squash each nodule separately with a sterile glass rod on a sterilised petri-dish with a small amount of sterilised water
- streak the fluid from the crushed nodules onto a pre-prepared YMA Congo Red plate with a sterilised inoculation loop (2-3 plates should be made with each nodule)
- number the plates and incubate at between 20-28°C

Observation

Check the plates regularly for <u>Rhizobium</u> growth. Fast growing colonies will appear within 2-4 days, whereas slow ones will take from 7-12 days.

Appendix 7. Preparation of media and solutions.

Yeast mannitol broth and agar (YMA)

Composition
- 10.0 g mannitol
- 0.2 g KH_2PO_4
- 0.8 g K_2HPO_4
- 0.2 g $MgSO_4.7H_2O$
- 0.1 g NaCl
- 1.0 g yeast extract
- 1000 ml distilled water

Preparation

To make the broth, mix the above ingredients and adjust to a pH of 7.2-7.4. If agar is required, 15.0 g/l of agar should be added and the mixture boiled for 10 minutes. To make slopes, dispense the medium in 10 ml aliquots into screw top culture tubes and autoclave. For plates, autoclave the boiled medium, cool and pour into petri dishes under sterile conditions.

YMA with Congo Red

Composition
- 2.5 ml Congo Red (1% W/V)
- 1000 ml YMA

Preparation

Sterilise the Congo Red solution separately and add to the YMA just before pouring the plates.

Glucose-Peptone Medium

Composition
- 5.0 g glucose
- 10.0 g peptone
- 15.0 g agar
- 1000 ml distilled water

Preparation

Melt the ingredients for five minutes at $115\,^{\circ}$C. If required, add 10 ml of Bromo-Cresol Purple indicator (1% in 95% ethanol) and then sterilise the solution for a further ten minutes at $121\,^{\circ}$C before plating.

Jensen's Agar Medium

Composition

- 10 ml of 20 g/l solution $\quad K_2HPO_4$
- 10 ml of 20 g/l solution either $MgSO_4.7H_2O$ or $NaCl$
- 10 ml of 100 g/l solution $\quad CaHPO_4$
- 10 ml of 10 g/l solution $\quad FeCl_3$

Preparation

To the mixed solution add 1 litre of distilled water and 12 g of agar (either Difco or Oxoid plain). Boil for 15 minutes, check and adjust pH to 6.8, pour into tubes (25 x 200 mm) and autoclave.

Nitrogen-free solution

Composition

- 0.4 g of 0.0023 M $\quad K_2HPO_4$
- 0.2 g of 0.0008 M $\quad MgSO_4.7H_2O$
- 0.8 g of 0.0058 M $\quad$ Anhydrous $CaSO_4$
- 1 ml of 5 mg/l Fe-chelate solution per litre of nutrient
- 1 ml of trace element stock solution per litre of nutrient as follows:

1.81 g/l $MnCl_2.H_2O$	2.03 g/l $MnSO_4.4H_2O$	
0.08 g/l $CuSO_4.5H_2O$	0.22 g/l $ZnSO_4.7H_2O$	
2.86 g/l H_3BO_3	0.025g/l $Na_2MoO_4.2H_2O$	
0.29g/l $CoSO_4.7H_2O$		

Preparation

Mix together the solutions, cover and leave to stand for a short while before use.

6. QUARANTINE AND SEED HEALTH OF FOOD LEGUMES

M.V. Reddy

Food Legume Improvement Program, ICARDA, Aleppo, Syria

Introduction

The transfer of plant materials across the globe could also spread associated diseases and pests and result in their widespread distribution. The benefits of a program of plant introduction and evaluation can be lost if pathogens and pests are also introduced. The plant health problems encountered in the transfer of food legume seed are pathological, nematological, entomological and parasitological.

Plant quarantine principles, methodology and suggested approaches

Principles

Plant quarantine regulations, made by a government or group of governments, restrict the entry of plants, plant products, soil, cultures of living organisms, as well as the packing materials, and the means of conveyance. The aim is to protect agriculture and the environment from avoidable damage by the inadvertant introduction of hazardous organisms.

Most plant quarantine regulations have the following features in common:

i.   Specify prohibitions.
ii.   Grant exceptions to prohibitions for scientific purposes.
iii. Require import permits.
iv.   Require phytosanitary certificates and/or certificates of origin.
v.   Stipulate inspection upon arrival.
vi.   Prescribe treatment upon arrival to eliminate a risk.
vii. Prescibe quarantine or post-entry quarantine isolation or other safeguards.

Organisms move from one region to another along natural pathways or along those created by man; quarantine is designed to prevent man from providing the means for the spread of pathogens. Many scientists subscribe to the hypothesis that all harmful organisms will eventually gain access to all regions of the world because quarantine can only delay their spread. However, thousands of hazardous organisms are still not distributed throughout the world, despite a long history of both human migration and trade in agricultural products. Moreover, agriculture will benefit from quarantine even if the spread of only some pathogens is delayed until resistance is incorporated into breeding lines.

Plant quarantine has two major roles to play in the distribution, and

J.R. Witcombe and W. Erskine (eds.) Genetic Resources and Their Exploitation - Chickpeas, Faba beans and Lentils. ISBN-13:978-94-009-6133-3(HB), 90-247-2940-8 (PB)
©1984, Martinus Nijhoff/Dr W. Junk Publishers for ICARDA and IBPGR.

maintenance of germplasm collections. The first involves the protection of agriculture from the inadvertant introduction of important pests and pathogens. The second involves the protection of the germplasm collections themselves from the ravages of domestic and foreign pests and pathogens.

Methodology
Plant quarantine regulations
Plant quarantine regulations control the following:

i.   The entry of potentially high risk genera by prohibition, post-entry quarantine, and restricted entry with permit.
ii.   The size, age and type of plant material of high risk genera.
iii. Alternate hosts.
iv.   Secondary or reservoir hosts.
v.   Packing materials.

Inspection and detection
Inspection of plant material for pests and pathogens involves specific methodology, some of which is outlined in Appendix 8. Usually inspection will be of seeds and often an incubation period is necessary. Inspection may involve the detection and identification of viruses; frequently virus like symptoms can be confused with genetic disorders and mineral deficiencies.

Treatments
i.   Heat treatment with hot water, hot air, or vapour heat. The principle is that plant material must be raised to a temperature that will kill the pest or pathogen without killing the plant or plant part.
ii.   Chemical treatment with sprays, dips, slurries, dusts or fumigants.
iii. Excision of diseased parts.
iv.   Heat therapy combined with meristem tip culture.

Utilizing quarantine facilities or isolation
i.   Quarantine stations/post-entry quarantine stations.
ii.   Geographic or climatic isolation.
iii. Isolation on premises of the importer.
iv.   Third country or intermediate quarantine.
v.   Consortium of plant quarantine stations.

Suggested approaches
The following suggestions are given to those involved in the exchange of germplasm:

i.   Evaluate the requirements for the introduction of germplasm.   (If possible, make an inventory of the genetic resources of the country or region).

ii.   Avoid the import of duplicate accessions.

iii. Channel requests for germplasm through a single plant introduction office.

iv.   Set up priorities for the country or region to ensure that quarantine facilities and services are utilized in high priority areas.

v.   Ensure that the distributor of germplasm is sending clean seed. Seed for distribution should be produced on Orobanche and nematode-free soil with a regular program of spraying against important insects and pathogens. It may require using an off-season growing facility.

Seed-borne fungi and bacteria

For seed-borne pathogens the procedures used in any quarantine program are the testing, screening, disinfection and decontamination of the seed; in emergency cases, healthy seed is produced by the isolation and cultivation of healthy tissues from infected seed or seedlings. The final, critical action is the release of the approved material.

Methods of detection

Different methods of seed health testing, mainly applicable for detection of fungi, are presented in Appendix 8. The tests could be of a generalised type, which may reveal a wide range of pathogens, or specialised, to detect a definite species, pathogenic race or a strain. The most commonly used generalised tests for detection of fungi are the ordinary ("standard") blotter test and the agar plate test, which reveal a wide range of fungi, mostly Imperfecti. Other, very versatile, testing procedures are seedling symptom tests and growing-on tests. The specialised tests are serological, and are used to detect and identify different viruses, or bacteria, or their strains; phage-plaque tests are used for the identification of strains of bacteria, and differential host tests are used for the identification of physiologic races of pathogens.

The screening of plant material destined for transfer from one region to another is quarantine policy in practice. The disease potential of pathogens must be weighed against the possible value of the seed to agricultural research.

Control measures

Seed treatment for the eradication of seed-borne pathogens can be used as an acceptable precaution in quarantine only if careful consideration is given to the limitations of such procedures. In general, seed treatment in agriculture may be regarded as successful even if only a reduction in inoculum is achieved.

In quarantine, no residual inoculum must remain after treatment.   The primary justification for seed-treatment as a quarantine precaution is its use as an additional safeguard to kill undetected trace amounts of inoculum in apparently healthy consignments.   Another acceptable policy would be to treat seeds which carry low amounts of inoculum in order to save particularly valuable material. It is unwise to accept for seed treatment consignments of seed that are badly infected or contaminated by organisms considered to be in strict quarantine, unless it is certain that the treatment is completely effective.

The following seed treatments may be considered:
- Heat treatment, including hot water treatment and aerated steam treatment.
- Organomercurials.
- Thiram, captan and systemic fungicides, among which benomyl and carboxin stand out.
- Insecticides, including aldrin and dialdrin, and fumigation with methyl bromide and phostoxin.

When there are only infected seeds of particularly valuable material, or systemically infected plants developed from these seeds, pathogen-free stocks may be obtained by meristem culture.

Seed-borne viruses

Methods of detection
i.   Visual examination of seed for externally-visible seed abnormalities associated with virus infection, such as seed-coat mottling in soybeans due to soybean mosaic virus infection.
ii.   Growing-on tests, in which the seedlings produced are examined for symptoms. This test, however, cannot detect symptomless viruses.
iii.   Infectivity tests, employing indicator plants that specifically react to certain viruses irrespective of symptoms in the tested material. This, when used along with growing-on tests, provides additional information.
iv.   Serological tests overcome the shortcomings of the above tests and are direct, specific, and rapid.
v.   Electron microscope tests.

Control measures
i.   Removal of infected seed.
ii.   Seed disinfection by seed cleaning, ageing, chemical and heat treatments, and irradiation (TMV on seed coats of tomato seeds was inactivated by treatment with 10% trisodium phosphate).
iii.   Seed certification.
iv.   Quarantine.

Seed-borne diseases of lentil, chickpea and faba bean

Lentil

The fungi that have been reported to be seed-borne in lentils include Alternaria alternata, Ascochyta lentis, Botrytis cinerea, Curvularia lunata, Helminthosporium spp., Aspergillus flavus, A. niger, Chaetomium spp., Macrophomina phaseolina, Thanatephorus cucumeris, Rhizopus spp., Corticium rolfsii, Fusarium moniliforme, F. semitectum, F. solani, F. oxysporum, Penicillium spp., Phoma spp. and Stachybotrys spp. (Khare, 1981). Species of Fusarium, Rhizoctonia, Aspergillus, Sclerotium and Botrytis are responsible for seed rot and seedling damage at the pre- and post-emergence stages.

Seed-borne F. oxysporum is responsible for the early wilting and drying of seedlings. Seed-borne F. roseum causes root-rot in the USA.

The sclerotia of S. sclerotiorum, and plant parts infected with Uromyces fabae or Peronospora lentis may propagate the disease when mixed with seed.

Among the viruses infecting lentils, pea mosaic is known to be seed-borne.

Seed treatment with fungicides is essential to control seed-borne pathogens. Seed treatment with Thiram + Brassicol or Thiram + Bavistin (1:1), 2.5 g/kg of seed or Benlate, 3 g/kg of seed reduced the incidence of wilt caused by F. oxysporum. Seed treatment with Thiram + Brassicol (1:1), at 2.5 g/kg of seed, has given promising results for collar rot caused by Corticium rolfsii. Brassicol, at 2 g/kg of seed has been reported to be effective for root rot pathogens. Seed treatment with Captan or Thiram, 2 g/kg of seed, has been reported to be effective for stem rot. Seed treatment with organomercurials has been recommended to eliminate seed-borne inoculum of rust caused by Uromyces fabae.

Chickpea

The fungi that have been isolated from chickpea seeds are Alternaria spp., Ascochyta rabiei, Aspergillus spp., Botrytis cinerea, Curvularia spp., Fusarium spp., Fusarium oxysporum, Penicillium spp., Phoma sorghina, Rhizoctonia bataticola and Rhizopus spp. Of these, A. rabiei, Alternaria spp., B. cinerea, and F. oxysporum are only isolated from surface-sterilised seeds. A. rabiei, B. cinerea and F. oxysporum are internally seed-borne (ICRISAT, 1977).

Seed treatment with Benlate and Calixin-M mixture (1:1), 3 g/kg of seed, or Calixin M, 3 g/kg, was found to completely eliminate A. rabiei from infected seed (Reddy, 1980). Seed treatment with Benlate T (Benomyl 30% and Thiram 30%), at 1.5 g/kg of seed, eradicated Fusarium completely, and almost all the other fungi mentioned above (Haware et al, 1978).

Faba bean

The important seed-borne fungal diseases of faba bean are chocolate spot (Botrytis fabae), and ascochyta blight (Ascochyta fabae). Among viruses, broad bean mottle virus (BBMV), Sudanese broad bean mosaic virus (SBBMV), broad bean stain virus (BBSV), bean yellow mosaic virus (BYMV), alfalfa mosaic virus (AMV), cucumber mosaic virus (CMV), broad bean true mosaic virus, pea early browning virus, pea seed-borne mosaic virus, and Vicia cryptic virus are known to be seed-borne.

Insect pests

When plant material is introduced into a country, the plant quarantine service has the responsibility of ensuring that the plants are adequately inspected and, if necessary, treated to prevent the introduction of insect pests. When conventional methods (e.g. fumigation) cannot be used to kill pests, since they reduce the viability of the germplasm, then other methods must be considered. These include cold storage for a period long enough to cause insect or mite death. Temperatures slightly below freezing may only arrest insect and mite development, and other techniques must be explored; growing the plant under protected, escape-proof environments, until the pest becomes exposed and susceptible to ordinary control methods, may be the solution.

In the conservation of germplasm, pests which ordinarily affect stored plant products may become a nuisance, or endanger the viability of the plant material. Stored seeds are subject to a number of pest problems, particularly beetles and moths, but insect-proof containers and conventional sanitation and protective measures are enough to keep pests away. In any case, seeds stored at low temperature and low relative humidity are not attacked by insects.

The bruchids known to infest faba bean seed are Bruchus dentipes, B. pisorum, B. rufimanus, Bruchidius albopictus, B. incarnatus and Callosobruchus chinensis. The bruchids infesting lentil seed are Bruchus ervi, B. lentis, B. rufimanus, B. dentipes, B. signaticornis, B. tristes, B. ulicus, Bruchidius minutus, Callosobruchus chinensis, C. analis, and C. maculatus (G.Hariri, 1981). In chickpea, Bruchidius albopictus, B. incarnatus, B. tocosus, Callosobruchus chinensis and C. maculatus are known to infest seed. Fumigation with methyl bromide and phostoxin are effective in controlling these bruchids. X-ray and Ashman infestation detectors for finding a hidden infestation of bruchids are also used.

Nematodes

In the past, some plant parasitic nematodes have been dispersed, in plants and plant products, from their native habitats to new areas. For example, Heterodera rostochiensis, the golden nematode of potato, spread into Japan, New Zealand and India. The small size of nematodes makes the effective implementation of quarantine regulations difficult. The quarantine measures against nematodes have to be applied indirectly, through restrictions on the importation of their known hosts, rather than direct interception of the infested material. Information on their mode of dispersal and their chances of establishment upon introduction is invaluable in devising effective quarantine regulations.

The introduction of nematodes over long distances has been mainly through soil attached to plant parts of different kinds used in germplasm exchange, and the most essential precautions are against the introduction of infested soil. Spread through the infested seed (e.g. Ditylenchus dipsaci on faba beans) and plant parts is another major means of dispersal, and the adoption of appropriate quarantine regulations and seed-health testing procedures, as described for fungi, is necessary.

Root-knot (Meloidogyne) and cyst (Heterodera) nematodes are known to infest faba bean, lentil and chickpea. The most serious problem, however, is stem nematode (Ditylenchus dipsaci) of faba bean. For this nematode, growing seed under nematode-free conditions with treatment with Temik and/or fumigation with methyl bromide is recommended.

Higher plant parasites

Broomrape (Orobanche spp.) is a serious parasitic weed of faba bean and lentil, and is also known to infest chickpea. It is propagated by small egg-shaped seeds which are 0.25 mm long and 0.15 mm thick. Cuscuta campestris is another parasitic weed of chickpea and lentil. Extreme care is needed to prevent the introduction of seed of these parasitic weeds with germplasm.

References

Hariri, G. 1981. Insects and other pests. Pages 173-190 in Lentils, eds. C. Webb and G. Hawtin, Slough, U.K: Commonwealth Agricultural Bureaux.

Haware, M.P., Nene, Y.L. and Rejeswari, R. 1978. Eradication of Fusarium oxysporum f. sp. cicei transmitted in chickpea seed. Phytopathology 68: 1364 - 1367.

ICRISAT, 1977. Pulse Pathology Annual Report. (1976 - 1977). Hyderabad, India: ICRISAT.

Khare, M.N. 1981. Diseases of lentils. Pages 163 - 172 in Lentils, eds. C. Webb and G. Hawtin, Slough, U.K.: Commonwealth Agricultural Bureaux.

Neergaard, P. 1977. Methods for detection and control of seed-borne fungi and bacteria. Pages 33-38 in Plant Health and Quarantine in the International Transfer of Genetic Resources, eds. Williams B. Hewitt and Huigi Chiarappa, Cleveland, USA: CRC Press.

Reddy, M.V. 1980. Calixin M - an effective fungicide for eradication of Ascochyta rabiei. International Chickpea Newsletter 3:12.p

Appendix 8. Seed-health testing procedures, summarised from Neergaard (1977).

i. Direct inspection

Examination of the dry seed, with impurities, using a hand lens or, preferably, a stereoscopic microscope. Seeds may be submerged in water to release spores and facilitate detection.

Application: Sclerotia of fungi, smut ball, nematodes on seeds, nematode galls, infected plant debris, e.g. Sclerotinia sclerotiorum, Botrytis cinerea, Claviceps purpurea, Ditylenchus dipsaci. Seeds discoloured or with lesions produced by fungi, bacteria, or viruses, e.g. ascochyta blight in chickpea, faba beans and lentil, viruses such as soybean mosaic in leguminous seed.

ii. Examination of suspension from washings of seed

An electro-mechanical shaker can be used to obtain standardised washings. Samples of the suspension are examined under a compound microscope.

Application: Covered smuts, e.g. Ustilago hordei and Tilletia spp. in cereals, oospores of certain downy mildews, ascochyta blight of chickpea.

iii. Whole embryo count method

Soaking grains overnight in 10% NaOH at 22°C, then washing with warm water through sieves of decreasing mesh size. Embryos finally cleared in lactophenol.

Application: Loose smuts of barley, Ustilago nuda, and wheat, Ustilago tritici and Ascochyta rabiei and Fusarium oxysporum of chickpea.

iv. Blotter method

Seeds are incubated on water-moistened blotter, usually for 7 days at 20°C. Sporulation of fungi is stimulated by near-ultraviolet (NUV) irradiation and a standard 12/12 h light/dark cycle. Petri dishes are usually used as containers. To allow penetration of the NUV, plastic or pyrex glass containers should be used. Sometimes blotters are soaked in 0.1-0.2% 2,4-D solution to counteract seed germination, thus aiding recording.

Application: The method used most commonly for detecting a considerable range of Fungi Imperfecti, including different spp. of Alternaria, Ascochyta, Botrytis, Cercospora, Colletotrichum, Diplodia, Drechslera, Fusarium, Macrophomina, Myrothecium, Phoma, Phomopsis, Septoria, and others, and for practically all kinds of seed.

v. Agar plate method

The seeds are plated in petri dishes on nutrient agar, in particular malt extract agar and potato-dextrose agar. Some selective media are available for specific tests. Light treatment as for the blotter test. Incubation for 5-7 days. Experienced analysts are capable of recording results by naked-eye examination, using colony characters as criteria.

Application: The classical procedure for Ascochyta spp. on pea, and for other fungi and hosts. Although slow growing fungi cannot be adequately detected, the procedure is relatively sensitive for revealing minor amounts of inoculum.

vi. Freezing method

A modified blotter method. After 1-2 days at $10\text{-}20\,^{\circ}\text{C}$, according to specification, incubation for some hours or for 1 day at $-20\,^{\circ}\text{C}$, then at $20\,^{\circ}\text{C}$ in NUV light for 5-7 days.

Application: Sometimes preferred for detecting certain fungi.

vii. Ordinary seedling symptom test

The seed is sown in autoclaved soil, sand, or similar material, and placed under normal daylight conditions for the observation of symptoms. A special procedure, the classical Hilter test, offers standard conditions for detection of seedling pathogens.

Application: Often used for detecting seedling symptoms which reveal pathogens rather than fungi to be identified, e.g. ascochyta blight of faba bean and chickpea.

viii. Water-agar seedling symptom test

The seeds are sown in water-agar, in 16 mm test tubes, one seed per tube, or in microculture plastic plates or petri dishes. They are placed under daylight conditions, e.g. 12/12/h cycle of artificial daylight and darkness. Seedlings are inspected for symptoms, and healthy seedlings may be transplanted for further post-quarantine cultivation.

Application: Can be used for many kinds of seed; an economical procedure which, in test tubes, ensures separation of healthy and infected seedlings. Has been used for detecting Drechslera graminea, D. sorokiniana, and D. teres in barley, Septoria nodorum and D. sorokiniana in wheat, and has been used for other pathogens and hosts.

ix. Indicator test, inoculation methods

A standard technique for identification of viruses, but also used for detection of trace amounts of pathogenic bacteria, e.g. by hypodermal injection of indicator plants with material from seeds slightly infected by the pathogen under test.

Application: Used for detection of Xanthomonas phaseoli and other bacteria in Phaseolus vulgaris, and Xanthomonas campestris in crucifers.

x. **Phage-plaque method**

Maceration of the seed to be tested, followed by incubation for 24 h to enable multiplication of bacteria. Samples of this material are transferred to sterile flasks, and a standard suspension of phage particles is added. Samples of these mixtures are plated immediately, and after 6-12 h, on plates with the indicator bacterium. Presence of homologous bacteria is indicated by a significant increase in the number of phage particles in the second plating.

Application: Used for detection of Pseudomonas phaseolicola and Xanthomonas phaseoli in Phaseolus vulgaris, Pseudomonas pisi in pea. Used for routine testing of foundation seed of Phaseolus vulgaris in Canada.

xi. **Serological methods**

Used to detect different seed-borne viruses (see Neergard, 1977 for methods).

7. UTILIZATION OF WILD RELATIVES AND PRIMITIVE FORMS OF FOOD LEGUMES

J.I. Cubero

Department of Genetics, Escuela Tecnica Superior de Ingenieros Agronomos,
Cordoba, Spain

The wild relatives

It has been remarked many times that we really know very little about the ancestors of our cultivated species. I will quote two statements by Harlan (1975):

i. 'In almost no crop do we have a large collection of wild relatives available in anticipation of need'.

ii. 'The study of the origin and evolution of cultivated plants is not only a work of intellectual and academic interest to the evolutionist, but a matter of enormous practical urgency in those crops that carry the burden of supporting the human population'.

To know the ancestral wild species of our crops is not, indeed, an easy task. This knowledge has to be achieved by means of a multidisciplinary approach; phytogeography, genetics, cytogenetics, biochemistry, archaeology, ethnology... to give an incomplete list. With an integrated methodology we can often determine the ancestors of our crops, but for many of the major crops the evidence is still incomplete. Surprisingly, we know the wild relatives of several unimportant crops because we are actually watching their domestication. Generally, the greater the area of cultivation of a crop, the greater is the distance to its wild relatives, and hence the more questions about its origin.

In food legumes, we find that all the cases are possible: wild relatives unknown, well known, and hypothetical. The known, or possible, wild ancestors of some food legumes are shown in Table 6.

It is obvious that if we are interested in the wild relatives of our crops, we have to tackle problems of taxonomy and systematics. Plant breeders require taxonomic systems with information about the possibility of transferring useful genes to the crop from the wild species. A very useful system is that of Harlan and de Wet (1971), which recognises three different gene pools according to the difficulty in transfer between them (Figure 4).

The primary gene pool (GP - 1) corresponds to the traditional concept of a biological species; cultivated and wild races are included in subspecies A and subspecies B, respectively. The number of subspecies in A is generally one. Several authentic B subspecies can be defined, their relationships with the cultivated one (or ones) ranging from very close to rather far, but always within the limit that they easily intercross with A to produce fertile hybrids

J.R. Witcombe and W. Erskine (eds.) Genetic Resources and Their Exploitation - Chickpeas, Faba beans and Lentils. ISBN-13:978-94-009-6133-3 (PB)
©1984, Martinus Nijhoff/Dr W. Junk Publishers for ICARDA and IBPGR.

Table 6. Food legumes and their possible wild ancestors.

Common Name	Botanical Name	Wild Relatives
Peanut	Arachis hypogea	A. monticola A. batizocoi (2x) A. cardenasii (2x)
Chickpea	Cicer arietinum	C. reticulatum
Soybean	Glycine max	G. soja
Lentil	Lens culinaris	L. orientalis, L. nigricans
Runner bean	Phaseolus coccineus	P. formosus, P. obvallatus
Lima bean	Phaseolus lunatus	P. lunatus
Common, haricot, French,dry,snap,bean	Phaseolus vulgaris	P. vulgaris, P.aborigineus
Pea	Pisum sativum	P. humile, P. elatius
Adzuki bean	Vigna angularis	V. umbellata, V.radiata, V. sublobata
Black gram, Urd bean	Vigna mungo	V. sublobata
Green gram, Mung bean	Vigna radiata Vigna aconitifolia	V. trilobata

Cowpea
  Cowpea
  Catjang bean
  Yard long bean
	Vigna unguiculata
	ssp. unguiculata
	ssp. cylindrica
	ssp. sesquipedalis
		V. unguiculata ssp.mensensis
		V. unguiculata ssp.dekindtiana

Bambara groundnut	Voandzeia subterranea	V. subterranea ssp.spontanea

which have good chromosome pairing, and normal gene segregation. Thus, any cross between forms belonging to this first gene pool must be considered intraspecific.

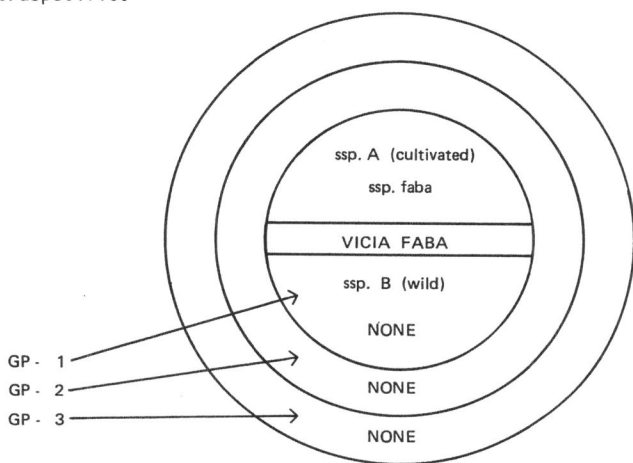

Figure 4. Vicia faba gene pools according to the scheme of Harlan and de Wet (1971).

The secondary gene pool (GP-2) includes the biological species that will cross with the species under consideration. In Harlan and de Wet's words, 'gene transfer is possible, but one must struggle with those barriers that separate biological species'. In spite of the high sterility level of the hybrids, poor (if any) chromosome pairing, and poor development of hybrids, both interspecific crosses and gene transfer are possible without special techniques, such as embryo culture. The allocation of a species to the primary or to the secondary gene pool can be rather difficult in some groups.

The tertiary gene pool (GP-3) includes species that can cross with that under study, but with the assistance of special techniques e.g. embryo culture, via artificially produced tetraploids, grafting, tissue culture, and the use of bridging species.

The origin and evolution of crops: Primitive populations vs. modern cultivars

The first farmers selected the first crops unconsciously, since harvesting and sowing the harvested seed exerts a selection pressure strong enough to result in the accumulation of useful mutations.

Crops change their gene pools by successive cycles of hybridisation with wild and weedy stocks.   Some strains are produced which retain many wild characteristics and are adapted to environments disturbed by humans - these are the companion weeds. Frequently these companion weeds follow the crop. In the new areas of cultivation new genes are absorbed by the crop, and new companion weeds are produced again if there are species or forms able to cross with the cultivated ones.

Even in the absence of active selection by farmers, this mechanism would produce a 'patchy' pattern of variability of the cultivated species throughout its growing area. This pattern will be unconsciously reinforced by selection by man, particularly if the species can be used in several ways. Hence, crops with a large cultivated area will have a greater opportunity to show a more intense differentiation. Crops in large areas are those which have the maximum interest to man.

In many cases, the taxonomic limits between wild relatives (including companion weeds) and primitive cultivated stocks are difficult to trace. A good example is Vicia sativa, the common vetch, where there is a continuous spectrum of variation from wild to cultivated forms, so that the wild forms are included within V. sativa, as subspecies or botanical varieties, in some taxonomic treatments.

The study of several cases reveals that 'primitive' man had achieved very precise selections for very specific purposes, and these had a very specific adaptation.   Some good examples are: Vigna unguiculata in Africa, with adaptation to photoperiod; Psophocarpus tetragonolobus with its several uses, from tubers to seeds; Vicia faba, with many forms such as the large-seeded varieties for direct human consumption.   Both modern and primitive populations of cultivated plants maintain a high level of genetic heterogeneity, very frequently more than the wild populations.   Among the characteristics of primitive populations are:

i.   The existence of highly adapted races; the formation of ecotypes can be very rapid (for example, peas in south America, common beans in Turkey). In particular, primitive populations are highly adapted to co-exist with parasites (disease, insects etc.)

ii.   The existence of 'genetic blocks' built up of highly coadapted genes.

iii.   Their evolution under the selection pressures created by cultivation.

Modern cultivars are highly uniform from a genetical point of view. Even in allogamous species the favourite method of breeders has been to produce uniform F_1 hybrids, and when this has not been possible they have produced synthetic varieties.   But even more important than genetic homogeneity is the very restricted genetic base of most of the modern cultivars.   In many cases, they are based on no more than two or three genotypes. Among the most important characteristics of modern cultivars are:

i. Their adaptation to 'sophisticated' environments, that is, highly modified by man.

ii. Stability in response to environmental differences between regions (i.e. they tend to have a low genotype-location interaction).

iii. Their evolution under the selection pressures created by breeders.

The use of wild and primitive populations

Transfer of characters

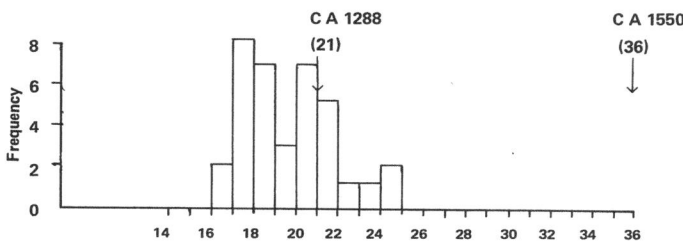

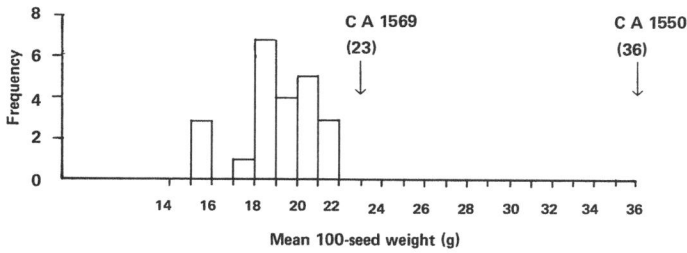

Figure 5. Distribution of F_2 populations and their parents in chickpea in two different crosses (C A numbers are accession numbers).

The use of both wild and primitive populations in plant breeding has been successful mainly in transferring qualitative characteristics such as resistance to diseases. The possibility of transferring quantitative ones is much more difficult because of the existence of highly coadapted genes forming 'blocks' which are difficult to break up. Sometimes, the polygenic system controlling a quantitative difference is so complex, since linkage, dominance and epistasis all have to be considered, that parental characteristics are not always easily recovered. For example, crosses between large and small seeded varieties of chickpea segregate in such a way that the large type is not easily recovered in the progenies (Figure 5, above). A program embarking on such crosses would either have large F_2 populations to increase the probability of recovering these forms or to consider backcrosses to the desirable parent.

For these reasons, the genes usually transferred from wild and primitive accessions have been 'qualitative' ones - resistance to diseases is typical. Resistance genes are easily detected and selected for, and all plant breeding texts have examples of such uses, e.g. modern cultivars of wheat, tomato and potato have resistance genes derived from wild species and primitive forms.

Widening the genetic base

Another use of these primitive materials in plant breeding is their use to enrich the impoverished gene pool of modern cultivars (see Zeven and van Harten, 1979). This genetic poverty has been very well established. For example, in the USA two cultivars of pea, out of 50, occupy 96% of the total area. Also, two varieties of dry beans represent 60% of the total acreage; and three out of 70 snap bean cultivars are sown in nearly 80% of the total area.

Moreover, most of the modern cultivars of the most widely grown crops have common ancestors. For example, most of the soybean cultivars grown nowadays in the USA are related to only eleven introductions, even though more than six thousand accessions were introduced between 1911 and 1930. Almost all the cultivars of peas available in the USA have 'Davis Perfection' or 'Alaska' as ancestors. Primitive forms, and, to a lesser extent, wild forms, can be successfully used to increase the gene pool of modern cultivars in medium or long term programs: good examples are groundnuts, common beans and faba beans (not wild strains in this case). Obviously, this enriched genetic pool will include both qualitative and quantitative characteristics. The latter are especially interesting because of the difficulties in transferring them in short term programs.

Direct selection

A third use of both wild and primitive forms is their direct use through selection. This use is more important in species not yet totally domesticated, or showing a continuum between primitive and modern forms (a large number of legumes are included here). But even in crops intensively worked on by breeders this possibility has been reported as useful. For example, in the potato a new population, neo-tuberosum, was selected from andigena crosses. However, the final use of such new materials in 'sophisticated' crops is their hybridisation with modern cultivars. The possibility of selecting new cultivars directly from primitive ones is mostly restricted to those species with only an incomplete degree of domestication, e.g. lupins, cowpeas, winged beans and bitter vetch.

However, this technique has been used in chickpea, faba bean and lentil (see below).

Production of new species

In the case of using wild species, the production of a new (synthetic) species is also a possible objective. This is performed by means of duplicating the chromosomic complement of the interspecific hybrid, provided that the parental species are phylogenetically far enough apart to show a lack of chromosome pairing in the hybrid. This will ensure a good pairing in the alloploid. Nature has performed this process in many genera: Triticum, Gossypium and Nicotiana are well known examples. Triticale is perhaps the best known example of a man-made crop. Other alloploids are used as bridge species. Among legumes of economic importance, Arachis is the best studied case. Groundnut breeders are extensively using artificially produced similar forms to be crossed with commercial varieties.

How to use wild and primitive forms

As has been seen, we can use these materials to :
- obtain new cultivars by selection from primitive forms.
- introduce qualitative characters into crops in short or medium term programs, and to introduce mainly quantitative characters in medium or long term programs.
- increase the gene pool of a species.
- produce new species (alloploids) to be used directly or used as 'bridge' species.

In other words, the three ways of using such materials are: selection, recombination and alloploidy. Selection does not produce any special problem, as it does not involve crosses. Crosses are necessary in the two other methods and, if we are dealing with wild forms, we will most probably have problems of sterility in obtaining hybrids.

It is worth mentioning that both intraspecific and interspecific crosses can depend on the genotype used as a parent. The experience obtained in the Triticineae demonstrates this well. The direction of the cross is also an important factor to consider. In legumes, there are several well documented cases in Indian Vigna and Phaseolus. For example, P. coccineus and P. vulgaris do not cross when cultivated materials are used and P. coccineus is chosen as the female parent, but they do cross when wild populations are used or, irrespective of using wild or cultivated stocks, when P. vulgaris is taken as the female parent.

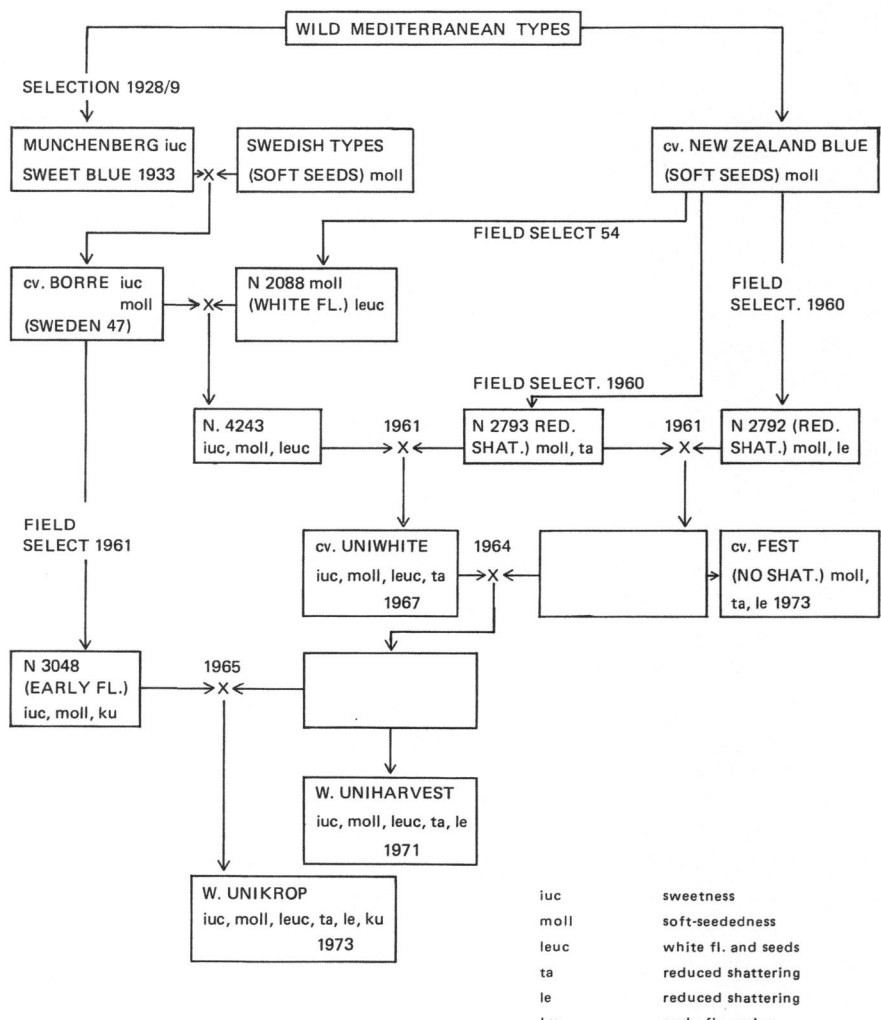

Figure 6. Origins of two cultivars of blue lupins (After Gladstones, 1970).

Some examples in legumes

Selection from primitive forms can produce excellent results in legumes (see Elsayed, Erskine, and Singh and Malhotra, Chapters 15, 18 and 11, respectively). However, the selection of recombinants after crossing is the most widely used way of exploiting wild and primitive forms, particularly if the yield of the cultivated species has been substantially increased by plant breeders.

Many important characteristics, such as resistance to disease and insects, winter hardiness, and adaptation to poor soils, are frequently lacking in the cultivated forms but are present in wild relatives. Primitive cultivars can also be the source of such characters, as well as for traits such as pod or seed type. A good example is the use of the landrace of Arachis hypogaea 'Tarapoto' to transfer resistance to Puccinia arachis to cultivated forms.

The breeding of two blue lupin cultivars started from wild Mediterranean types in 1928 (Figure 6). Of course, most of the mutants seen in Figure 6 were not found in authentic wild and/or primitive forms, but on materials already selected. But as 'Unicrop' lupin has been developed in less than fifty years from wild origins, we can suppose that most of these useful mutations were already present in the wild material.

Since wild and primitive forms can be used to increase the gene pool of a cultivated species we are crossing different faba bean cultivars with primitive paucijuga accessions. We are looking for new forms that combine in different ways self-fertility, seed size, branching, etc. The new forms obtained will be used in hybridisation programs rather than as new cultivars, but, in principle, their direct use is not excluded. Crosses with primitive and wild forms of Phaseolus coccineus have also been reported and groundnut breeders are also using this way of increasing variability. Resistance to several diseases was successfully transferred to Vicia sativa by crossing it with V. angustifolia. The method was to use the hybrids as donors in a backcross program, and also to select amongst the descendents of the hybrids for resistant lines, which were used in a further step as donors. But in some cases, the production of interspecific hybrids has proved impossible. This is the case in faba beans. Many groups have tried to cross them with wild Vicia species, without any success up to the present time.

Alloploidy has only been successfully exploited in the groundnut, the most important (from an economic viewpoint) polyploid legume (Figure 7). Plans to obtain alloploids as 'bridge species' for the transfer of useful characters from wild to cultivated stocks have been devised for faba beans; but they are only written and not yet achieved. Most of the cultivated legumes do not need, in principle, such methodology, either because they have been intensively worked upon (peas, common beans, soyabeans) or because, more commonly, selection and intraspecific crosses can provide most of the materials required.

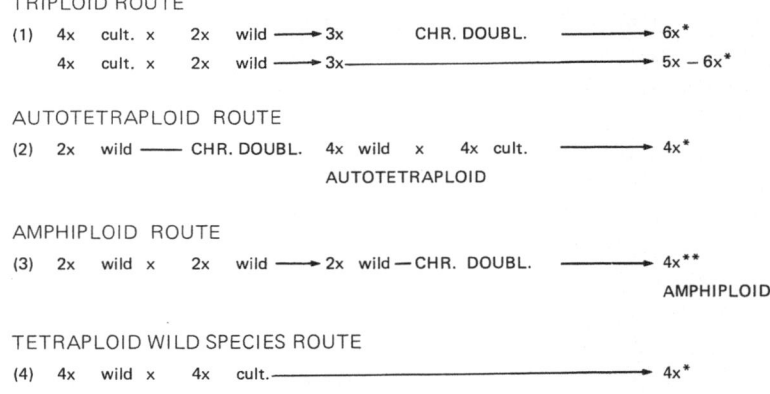

Figure 7.   Utilisation of wild Arachis (After Singh, Shastri & Moss, 1977).

Status of research on cultivated legumes

We will include in the expression 'food legume' all legumes used for food, even soyabeans and peanuts, which are generally considered as oil crops, and 'green' peas and 'green' common beans, which are not normally considered as a staple food.

According to the number of studies performed, and the level of knowledge reached, food legumes can be divided into three groups:

Species which have been intensively studied

Large collections of all kinds of material have been formed (modern cultivars, primitive forms, wild relatives) and evaluated.   Collections are continually growing, but this increase is now rather slow.   Breeding has resulted in high yielding cultivars, and in the near future sophisticated methods will have to be used to improve on their performance.   The possibility of extra-specific genetic transfer has been studied.   There have been many studies on quality, and its genetic control is well understood.   Interspecific crosses have been made and they have been used in theoretical and practical studies. Even if the crosses have proved to be very difficult or impossible, this knowledge has been obtained after long and intensive studies.

Obviously, peas, Phaseolus beans, soyabeans and peanuts belong here.

Species with a shorter (or a less intense) research history

These are valuable crops which are presently under-researched. Collections are being actively formed, and they are expanding rather quickly, but there are geographical areas yet to be explored. There are several, but not many, groups working on them. (This does not refer to special studies: V. faba is good material in cytogenetics but not because of the interest placed in V. faba). Classical methods of breeding are still the most useful, with a maximum priority on hybridisation. Interspecific crosses have been tried, but the studies are far from complete. The number of studies on quality are increasing. Knowledge of their genetics is, in general, rather poor.

We find in this group authentic 'food legumes'. We can include here faba beans, cowpeas, some Phaseolus, Mediterranean lupins as a group, and perhaps asiatic Vigna also as a group. Chickpea is quickly reaching this status, but it would be better placed in the next group.

Species poorly known

Collections are being formed, but the total knowledge about them is scant and there are many 'blanks'.

Most of the varieties are landraces or 'primitive'. Classical methods of breeding are the most fruitful, and very simple methods (mass selection, plant - to - line, etc.) are advisable in most regions. Knowledge of their genetics is very poor or non-existent. Generally speaking, they are grown in restricted areas, mostly in developing countries. Very few researchers, if any, are working on them. All the species have a short research history.

Most of the food legumes belong to this group. Some of them can reach the status of the second group in a rather short time: lentils, tarwi (Lupinus mutabilis) and other lupins, winged bean, some Vigna. But for Lathyrus, Voandzeia and Vicia spp. grown for forage, the work is yet to be done.

References

Gladstones, J.S. 1970. Lupins as crop plants. Field Crop Abstracts 23: 123-148.

Harlan, J.R. 1975. Crops and Man. Madison: American Society of Agronomy.

Harlan, J.R. and de Wet, J.M.J. 1971. Towards a rational classification of cultivated plants. Taxon 20: 509-517.

Zeven, A.C. and van Harten, A.M. 1979. Broadening the Genetic Base of Crops. Wageningen: Centre for Agricultural Publishing and Documentation.

Further reading

Frankel, O.H. and Bennett, E. (eds.) 1970. Genetic Resources in Plants - their Exploration and Conservation. Oxford: Blackwell.

Milner, M. (ed.) 1975. Nutritional Improvement of Food Legumes by Breeding. Washington, D.C.:John Wiley.

National Academy of Science 1972. Genetic Vulnerability of Major Crops. Washington, D.C.: NAS.

Razdan, M.K. and Cocking, E.C. 1981. Improvement of legumes by exploring extra-specific genetic variation. Euphytica 30: 819-833.

Singh, A.H., Sastri, D.C. and Moss, J.P. 1980. Utilization of wild *Arachis* species at ICRISAT. Pages 87-93 in Proceedings, International Workshop on Groundnut, ICRISAT, 13-17 October 1980, Patancheru, India: ICRISAT.

Smartt, V. 1980. Evolution and Evolutionary Problems in Food Legumes. Economic Botany 34: 219-235.

Summerfield, R.J. and Bunting, A.H. 1980. (eds.). Advances in Legume Science. Kew: Royal Botanical Gardens.

Ucko, P.J. and Dimbleby, G.W. 1969. The Domestication and Exploitation of Plants and Animals. London: Gerald Duckworth and Co.

8. UTILIZATION OF GENETIC RESOURCES IN A NATIONAL FOOD LEGUME PROGRAM

Nasri Haddad

University of Jordan, Amman, Jordan

Local and international food legume collections

Local collections of landraces are a readily available source of genetic material for any national program. The local collection usually contains material which is highly adapted to the local environmental conditions and has acceptable seed characters. Therefore, national programs should have a complete collection which represents the whole of the country. Local material should be characterized and evaluated at different locations within the country.

Material in the local collection (and that introduced from outside the country which shows good adaptability to the local environment) is the main genetic resource for crop improvement by direct selection. It can also be utilized by selecting genotypes with specific characters to be used as parental material in the hybridisation program.

Variability existing within the local and international collections of the three food legumes with which we are concerned is reported to be high. This variability indicates the genetic potential of these crops. Moreover, it is genetic material available to the national programs.

Lentils (Lens culinaris Med.)

International germplasm collections of Lens for breeding purposes are maintained in the United States, India, and at ICARDA, Syria. ICARDA maintains a collection (ILL) of about 5400 accessions from 56 different countries (Solh and Erskine, Chapter 17). However, there are also national local collections, and many of these are duplicated in the ICARDA collection.

Variation in national collections

How much of the genetic variability within these lentil collections can be utilized by national programs? The results of the few studies carried out on some local collections indicate the existence of considerable variability. In Turkey, Eser (1964) conducted a characterization study on 333 lentil lines obtained from different parts of Turkey, and found a considerable variation for some botanical and agronomic characteristics, such as plant height, number of leaflets per leaf, length of leaflet, number of flowers per inflorescence, flower length, pod length, pod width, and seed characteristics. In 1982, 152 locally collected, pure line, accessions of the University of Jordan Lentil (UJL) Collection were evaluated for different characteristics by the author.

J.R. Witcombe and W. Erskine (eds.) Genetic Resources and Their Exploitation - Chickpeas, Faba beans and Lentils. ISBN-13:978-94-009-6133-3 (PB)
©1984, Martinus Nijhoff/Dr W. Junk Publishers for ICARDA and IBPGR.

Variability among these lines was significant for all the characters studied. These characters included growth habit, plant height, height of the lowest pod, number of leaflets and width of leaflet, number of primary and secondary branches, grain and biological yield, 100-seed weight and seed coat colouration.

These studies indicate that considerable variability does exist in the local material. National programs should utilize the desirable aspects of this variability by collecting and evaluating their local genotypes and searching for desirable genes.

Exploiting the variation

The potential of local material of lentil can be observed from results obtained in Jordan and India. Promising lines with different origins received from ICARDA (Lentil International Screening Nursery-Small and Large; LISN-S and LISN-L) were evaluated in Jordan. The best five lines in the LISN-S were all selections from lines collected from Jordan, whereas the best five lines in the LISN-L originated in Syria, Iraq, Cyprus and Jordan (Haddad, 1982). This indicates the high adaptation of these lines to their local environment. Tiwari and Singh (1980) evaluated 925 entries which were collected from different parts of India, which included 47 old and new varieties and 34 exotic introductions from eleven countries. They identified 13 as excellent lines and recommended these to be used as parents in lentil breeding programs at Jabalpur, India. These 13 lines were all of Indian origin.

Lines carrying genes for disease and insect resistance have been identified in lentil collections. Immunity to pea seed-borne mosaic virus was found in the USDA lentil collections in the following accessions: P.I. 212610, 251786, 207745, and 368648, collected from Afghanistan, USSR, Greece and Yugoslavia, respectively (Haddad et al., 1978). Lines JI-500 and JI-764 from the India collection were found to be promising sources of resistance/tolerance to strains of wilt, Fusarium oxysporum f. sp. lentis (Kannaiyan et al. 1978). Pandya et al. (1980), in northern India, developed a new variety, Pant L406, which is resistant to rust and wilt diseases and is a selection from line P495.

Lentil collections carry a good genetic potential for other characters. A selection from the USDA, PI 179310 (known as PI 179310 TR), exhibited a high level of tolerance to the herbicide MCPB (Slinkard, 1980). Moreover, fifteen lines from the USDA collection were found to be salt tolerant (Jana and Slinkard, 1979). Screening a total of 3592 lentil accessions from the ICARDA collection for cold tolerance over a severe winter near Ankara, Turkey, revealed that 238 lines were cold tolerant (Erskine et al., 1981).

In spite of the fact that most of the area under lentil cultivation is planted to landraces (Solh and Erskine, Chapter 17), new improved cultivars are released and planted in some countries. The majority of these improved varieties are the result of direct selection from the local and introduced materials. Some of these are: Precoz in Argentina (Riva, 1975); Pusa-4, Pusa-6, Pusa-8 and Pusa-10 in India (Lal et al., 1976); Laird (Slinkard, 1978) and Eston (Slinkard, 1981) in Canada; and Red Chief (Muehlbauer and Wilson, 1981) in the United States.

Chickpeas (Cicer arietinum L.)

International chickpea collections are located at ICRISAT, and at ICARDA. The main purpose of these collections is to assist breeders in making more effective use of the available germplasm.

The ICRISAT collection consisted of 11000 accessions in 1979. Evaluation of part of the collection was carried out during three seasons at ICRISAT (ICRISAT, 1979). Very promising lines were reported; 81 accessions were found to be less susceptible to Heliothis armigera; 116 lines, out of the 3110 screened, scored 5 or less against ascochyta blight on a 9 point scale, with 1 as highly resistant; 131 lines were promising for stunt resistance, and six lines were promising against both wilt and ascochyta blight (ICC 299, 595, 599, 843, 4716 and 5006). Moreover, one line (ICC 8181) was found to be promising against both wilt and stunt. By 1980 the collection had increased to 12201 accessions (van der Maesen, 1980).

ICRISAT is establishing a Data Management and Retrieval System (IDMRS), to provide information (sub-sets of the accession catalogue) to assist users in selecting accessions. Therefore, national programs should utilize this system to determine the genetic stocks they need to use in their breeding work. However, it should be kept in mind that material showing good levels of performance under the conditions of the ICRISAT research station may not do so under the different environments of the local programs. Re-evaluation of these materials under local conditions is therefore necessary.

The ICARDA chickpea germplasm collection comprises 4420 kabuli accessions from 34 countries. Work on this collection is described by Singh and Malhotra (Chapters 10 and 11). This collection is a useful source of variability for national programs.

Faba beans (Vicia faba L.)

Considerable genetic variation has been reported within the species and this germplasm resource is still largely unexploited. The International Legume Faba bean (ILB) collection is maintained at ICARDA, and in 1980 had 1931 accessions from 50 countries (Hawtin and Omar, 1980). Some countries are poorly represented. There are other collections in such countries as the USSR, Italy, Germany (GDR and FRG), the Netherlands and Czechoslovakia (Witcombe, Chapter 13).

The maintenance of faba bean collections is complicated by the high level of out-crossing in this plant; after many generations of seed increase under open field conditions, the separate identity of each line is lost.   Seed increase should therefore be carried out in isolation plots, in insect-proof screen cages, or through bagging individual plants.   A working collection has been established at ICARDA by selecting from one to four single selfed plants in each accession and then, in the following season, growing their progenies in separate rows in a screen house.   The maintenance of a pure-line collection requires resources beyond those of most national programs, but breeders in the national programs can rely on the ICARDA collection as a source of material.

The genetic potential of faba bean pure line collections is demonstrated by the identification of genes for disease resistance; out of 1650 pure lines screened for Botrytis fabae (chocolate spot) resistance, 21 entries were rated one (without disease symptoms), and one selection from ILB 438 was rated highly resistant (Hanounik and Hawtin, 1982).   Resistance to rust, Uromyces fabae, was identified in 3 lines out of 25 screened.   These three lines consistently behaved as slow rusters (Bernier and Conner, 1982).   F 402, a selection from the Egyptian collection, was found to be resistant to Orobanche (Nassib et al., 1982).   The author believes that resistance genes to Orobanche could be found in Egyptian landraces.   Several lines from Egypt, Syria, and Sudan were found to be highly autofertile; a desirable character in faba bean (Hawtin, Chapter 14).

Hybridisation

Breeding lines and segregating populations
Selection of parents
The easiest way to develop a new variety in a national food legume program is by direct selection from the local landraces and from introduced material. This, in fact, is the oldest method practised by farmers and plant breeders. However, when national programs need to combine desirable characters from different sources, to fully exploit the available genetic resources, then hybridisation is necessary.

Plant breeders cross parents with complementary traits to develop improved cultivars, and to select breeding stocks for potential use as parents in their programs.   The choice of parents to be crossed is an important decision which will depend on the goals of the program and the breeding method used (Forsberg and Smith, 1980).   Parental material, especially in newly established national programs, is mainly derived from local collections or from introductions from areas with a similar environment.   This material should be adapted and may also have a good level of pest tolerance.   Direct contacts with other national programs that are breeding for similar environmental conditions should be established to obtain parental materials.   International genetic resources centres of food legumes, such as ICARDA and ICRISAT, will be of great help in providing such material upon request.

National programs should ask for material with characters which satisfy their objectives, since it is impracticable for them to evaluate and screen the world collection. The decision of what to choose from the world collection will be easier when the international genetic resources centres publish descriptor catalogues of their genetic material.

Regardless of the source of the genetic material, it is very important for each program to evaluate agronomic traits and disease resistance under the environmental conditions of the region where the germplasm will be used. Disease races usually differ from one area to another; thus multi-location tests should be conducted to confirm the resistance of the introduced material.

Crossing technique and breeding methods

Hybridisation in national programs needs skilful technicians, especially for lentils and chickpeas. The percentage of successful pollinations varies widely with individuals and environments. In lentil, Wilson (1972) reported that the success of manual pollination ranged from 46 to 82% under greenhouse conditions. However, in India, successful pollinations under field conditions were reported to range from 27 to 43% (Malhotra et al., 1978). With chickpeas, the percentage of successful pollination varies depending on environmental conditions. Seed set usually ranged from 10 to 20%, but reached 70 to 80% under certain controlled conditions (Auckland and van der Maesen, 1980). With faba bean, hybridisation is usually successful provided the requirements for full flower development are met (Bond et al., 1980). After considering the difficulties involved, each program should decide upon the stage at which it can start its own hybridisation activities.

Hybridisation becomes very important when disease problems are facing the national program. Therefore, identification of sources of resistance in the collections will be the first task for the program, followed by crossing this material with locally adapted parents. Backcrossing is usually used to transfer genes for resistance and to maintain the merits of the adapted parent. Nevertheless, when the disease resistance is complex, and when a recombination of more than one source of resistance is required, then other breeding methods should be considered.

Regardless of the breeding method used, selection of certain breeding lines with specific characters will be practised. These will be maintained as pure lines and will be added to the program's genetic resources. They will be utilized in the future, either as parental material in new crosses, or included in the screening nurseries or varietal trials.

ICARDA provides interested people with materials selected from their segregating populations. They also provide bulks of their segregating populations. These are very valuable genetic stocks which could be utilized by various national programs. It is advisable that early generations of their segregating populations, such as the F_2 and F_3, should be tested by the national programs. This permits the selection of genotypes which perform well under local environmental conditions and allows inferior populations to be discarded at an early stage.

Introgression

The value of crossing between divergent sub-groups within a species has been recognised by plant breeders for some time. Crossing between the two major sub-groups of chickpeas, desi and kabuli, was conducted at ICRISAT, ICARDA, and some other institutes. This procedure was successful, since each type benefited from the transfer of certain specific genes from the other. Hawtin and Singh (1979) reviewed this subject; they concluded that the kabuli type might be improved by the transfer of greater numbers of secondary branches or heat tolerance from the desi group, which in turn might benefit from the addition of genes for taller, more erect growth habit or cold tolerance from the kabulis. However, the introgression of a few specific genes from one group into the other can best be achieved by a conventional backcrossing program. National food legume programs should be aware of the genetic potentialities of such crosses, but a good understanding of the genetical, morphological, physiological and biochemical characteristics of the two sub-groups is essential.

Utilization of wild relatives and polyploidy

Hybridisation between the cultivated forms and their wild relatives is practised in national programs to:

i.   Transfer one or more genes from one species to another.

ii. Achieve a new character expression not found in either parent.

Wild hybridisation involves crossing a locally adapted parent with a wild parent (which will usually be unadapted). The purpose is to introduce a desired character (commonly, but not always, disease resistance) from the foreign parent. Subsequent backcrossing to the adapted parent maintains desirable agronomic characteristics in the resultant hybrids. Wild hybridisation has been tried with the food legumes but with limited success. In lentils, crosses were made between the wild species, Lens orientalis, and the cultivated species, Lens culinaris. Although the two species crossed readily, the usefulness of these crosses in lentil improvement is still not well established (Hawtin, 1979). In chickpeas, the wild species Cicer reticulatum appears to be a promising source of resistance genes, especially for ascochyta blight. Cicer reticulatum was crossed with G-130, JG-62 and P-5462 at ICRISAT to transfer ascochyta blight resistance to cultivated chickpeas. Other crosses, using other wild species, have been attempted with limited success (van der Maesen et al., 1979). As for faba beans, crossing between Vicia faba and other Vicia species has so far proved unsuccessful (Cubero, Chapter 12). However, if breeding barriers can be overcome there is a great potential for improvement by this method. For example, one of the closest relatives of the faba bean, Vicia narbonensis, a wild, weedy species found in the Middle East, has been shown to have resistance to a number of diseases and pests, including chocolate spot, ascochyta blight, and certain aphid species (Hawtin, 1979).

Hybridisation, followed by chromosome doubling, can be used to produce alloploid species. The possibilities in food legumes are certainly worth exploring, although results to date are not encouraging. Recently, autotetraploid chickpea was obtained at ICRISAT (Pundir et al., 1981). It has larger leaves, flowers, stomata, pollen grain and seeds, but the number of pollen grains produced and pod set were reduced. In Vicia faba, the first fertile tetraploid, and the first aneuploids have been described (Martin, 1979; Poulsen and Martin, 1977). The tetraploids exhibit agronomic disadvantages, such as late development, low tillering ability and poor growth under artificial conditions. It seems that the major problem in the breeding of autotetraploids, and the main limitation to their practical use as a crop, is partial sterility.

Mutation breeding

The use of mutations, resulting from irradiation or chemical mutagens, has not received much attention as a breeding method in the major food legumes. Mutation breeding has merits if the desired character is simply inherited, and if suitable screening techniques are available to identify that character in large populations. It is better suited for inducing recessive mutations rather than dominant ones. Some mutants have been induced in lentil, and almost all of these are morphological mutants with limited use. In chickpeas, a spontaneous polycarpellary mutant has been reported in ICRISAT (Pundir and van der Maesen, 1981). Mutation breeding in faba bean was reviewed by Abdalla (1982); different mutagenic agents had been used to induce variability for vegetative, generative and yield characters, as well as protein content. Other areas suggested for future mutation research include pest resistance, plant models, fertility and incompatibility, and resistance to environmental stresses.

Wild hybridisation, polyploidy and mutation have a potential in plant breeding, as a tool to create genetic variability and to introduce new genes. Food legume scientists should utilize this potential in their programs. However, there are many problems associated with their direct utilization, such as partial sterility and screening techniques. Research in these areas requires facilities and techniques which may not be available in national programs. Therefore, they should concentrate on the genetic resources which are readily available in collections and in segregating populations which can be directly utilized. However, national programs should be aware of research in other areas conducted by international programs and universities, and should exploit the results in their breeding work.

References

Abdalla, M.M.F. 1982. Mutation breeding in faba beans. Pages 83-90 $\underline{in}$ Faba Bean Improvement, eds. G. Hawtin, and C. Webb, The Hague: Martinus Nijhoff.

Auckland, A.K. and van der Maesen, L.J.G. 1980. Chickpeas. Pages 249-259 $\underline{in}$ Hybridisation of Crop Plants, eds. W. Fehr and H. Hadley, Madison, Wisconsin: American Society of Agronomy.

Bernier, C.C. and Conner, R.L. 1982. Breeding for resistance to faba bean rust. Pages 251-257 $\underline{in}$ Faba Bean Improvement, eds. G. Hawtin, and C. Webb, The Hague: Martinus Nijhoff.

Bond, D.A., Lawes, D.A. and Poulsen, M.H. 1980. Broad bean (Faba Bean). Pages 203-213 $\underline{in}$ Hybridisation of Crop Plants, eds. W. Fehr and H. Hadley. Madison, Wisconsin: American Society of Agronomy.

Erskine, W., Meyveci, K. and N. Izgin. 1981. Screening of a world lentil collection for cold tolerance. LENS 8: 5-9.

Eser, D. 1964. Turkiyelde Yetistirilen Mercimek Cesitlerinin Onemli Morfologik Karakterleri Uzerinde Arastirmalar: Ankara Universitesi; Turkey.

Forsberg, R.A. and Smith, R.R. 1980. Sources, maintenance and utilisation of parental material. Pages 65-81 $\underline{in}$ Hybridisation of Crop Plants, eds. W. Fehr and H. Hadley. Madison, Wisconsin: American Society of Agronomy.

Haddad, N. 1982. Food legume improvement project in collaboration with IDRC. Annual report 1981. Amman, Jordan: University of Jordan.

Haddad, N., Muehlbauer, F.J. and Hampton, R.O. 1978. Inheritance of resistance to Pea seed-borne mosaic virus in Lentils. Crop Science 18: 613-615.

Hanounik, S. and Hawtin, G.C. 1982. Breeding for resistance to chocolate spot caused by Botrytis fabae. Pages 243-250 in Faba Bean Improvement, eds. G. Hawtin and C. Webb, The Hague: Martinus Nijhoff.

Hawtin, G.C. 1979. Strategies for genetic improvement of lentils, broad beans, and chickpeas, with special emphasis on research at ICARDA. Pages 147-154 in Food Legume Improvement and Development, eds. G.C. Hawtin and G. Chancellor, Ottawa: ICARDA-IDRC.

Hawtin, G.C. and Omar, M. 1980. International faba bean germplasm collections at ICARDA. FABIS 2: 20-22.

Hawtin, G.C. and Singh, K.B. 1979. Kabuli-Desi introgression: problems and prospects. Pages 51-60 in Proceedings, International Workshop on Chickpea Improvement. ICRISAT, Feb-March 1979, Hyderabad, India: ICRISAT.

ICRISAT, 1979. ICRISAT's Chickpeas Genetic Resources. Introduction to Descriptors and Data Evaluation. ICRISAT, Genetic Resources Unit, Hyderabad, India: ICRISAT.

Jana, M.K. and Slinkard, A.E. 1979. Screening for salt tolerance in lentils. LENS 6: 25-27.

Kannaiyan, J., Nene, Y.L. and Pant, G.B. 1978. Strains of Fusarium oxysporum f. sp. lentis and their pathogenicity on some lentil lines. LENS 5: 8-10.

Lal, S., Chandra, N. and Tikoo, J.L. 1976. Some improved strains of lentil for India. LENS 3: 11-12.

Malhotra, R.S., Balyan, H.S. and Gupta, P.K. 1978. Crossing techniques in lentil. LENS 5: 7-8.

Martin, A. 1979. Aneuploidy in Vicia faba L. Journal of Heredity 69: 421-423.

Muehlbauer, F.J. and Wilson, V.E. 1981. Red Chief. A new large-seeded lentil cultivar. LENS 8: 31.

Nassib, A.M., Ibrahim, A. and Khalil, S.A. 1982. Breeding for resistance to Orobanche. Pages 199-206 in Faba Bean Improvement, eds. G. Hawtin and C. Webb, The Hague: Martinus Nijhoff.

Pandya, B.P., Pandey, M.P. and Singh, J.P. 1980. Development of Pant L406-lentil. Resistant to rust and wilt. LENS 7: 34-37.

Poulsen, M.H. and Martin, A. 1977. A reproductive tetraploid *Vicia faba*. Hereditas 87: 123-126.

Pundir, R.P.S. and van der Maesen, L.J.G. 1981. A spontaneous polycarpellary mutant in chickpeas (*Cicer arietinum* L.). International Chickpea Newsletter 5: 2-3.

Pundir, R.P.S., Roa, N.K. and van der Maesen, L.J.G. 1981. Induced autotetraploidy in chickpeas. International Chickpea Newsletter 4: 7-8.

Riva, A. 1975. Precoz, a new lentil cultivar for Argentina. LENS 2: 9-10.

Slinkard, A.E. 1978. Laird lentil licensed in Canada. LENS 5: 4.

Slinkard, A.E. 1980. Resistance to MCPB in P.I. 179310 TR. LENS 7: 65.

Slinkard, A.E. 1981. Eston, a new small-seeded lentil cultivar. LENS 8: 30-31.

Tiwari, A.S. and Singh, B.P. 1980. Evaluation of lentil germplasm. LENS 7: 20-22.

van der Maesen, L.J.G. 1980. Chickpea genetic resources at ICRISAT. International Chickpea Newsletter 3: 2-3.

van der Maesen, L.J.G., Pundir, R.P.S. and Remananden, P. 1979. The current status of chickpea germplasm work at ICRISAT. Pages 23-32 in Proceedings of the International Workshop on Chickpea Improvement. ICRISAT, Feb-March 1979, Hyderabad, India: ICRISAT.

Wilson, V.E. 1972. Morphology and techniques for crossing *Lens esculenta* Moench. Crop Science 12: 231-232.

9. TAXONOMY, DISTRIBUTION AND EVOLUTION OF THE CHICKPEA AND ITS WILD RELATIVES

L.J.G. van der Maesen
ICRISAT, Hyderabad, India

The genus Cicer L.

The genus Cicer, earlier classified in the tribe Vicieae of the legume family, along with vetches, lentils, and faba beans, is now classified in its own tribe, the Cicereae Alef.. Kupicha (1977) found that pollen morphology and vascular anatomy set Cicer species further apart from the other members of Vicieae (Lathyrus, Lens, Vicia, Pisum, and Vavilovia) and closer to the tribe Trifolieae, which differs from the Cicereae only in having hypogeal germination, tendrils, stipules free from the petiole and nonpapillate unicellular hairs.

van der Maesen (1972) recognised 39 species of Cicer, 31 of them perennial species and 8 annuals. Since then, one annual species has been described: C. reticulatum Ladizinsky (1975). Other species described recently, but not inspected by the author, are: C. heterophyllum Contandriopoulos et al. (1972), which is close to C. montbretii Jaub. & Spach; C. laetum Rassulova and Sharipova; and C. rassuloviae Linczevski. Appendix 9 gives an overview of the sections of the genus, their main characteristics, and the species contained in each section.

A good knowledge of the species of Cicer, especially of the chickpea, C. arietinum L., and its nearest annual wild relatives, is necessary to enable the scientific community to make use of the genetic resources needed for chickpea improvement. A germplasm collection of over 12000 accessions at ICRISAT and ICARDA constitutes a firm base with few gaps, although the wild species are still insufficiently represented in collections.

Distribution

Broadly speaking the species Cicer are adapted or confined to four types of environment:

i.   In cultivated environments: the chickpea (Cicer arietinum) is found only in cultivation, sometimes as an escape from cultivation or as a volunteer, and is unable to colonise successfully without intervention by man.

ii.   In weedy habitats (e.g. Cicer reticulatum, C. bijugum): such species grow in fallow or disturbed habitats, roadsides, cultivated fields of wheat, and other places completely untouched by man or cattle.

iii.   On mountain slopes among rubble (e.g. Cicer pungens, and C. yamashitae): apparently seed dispersal is less hampered by predation that in habitats

J.R. Witcombe and W. Erskine (eds.) Genetic Resources and Their Exploitation - Chickpeas, Faba beans and Lentils. ISBN-13:978-94-009-6133-3(PB)
©1984, Martinus Nijhoff/Dr W. Junk Publishers for ICARDA and IBPGR.

lacking stones, as the seeds are better protected between stones. The root system is a strong taproot, becoming woody with age in perennials.

iv. On forest soils, in broad-leaf or pine forests (e.g. C. montbretii, C. floribundum): The humus-rich layers are exploited but a deep taproot is also present. These species prefer some shade.

Cicer spp. occur from sea level (e.g. C. arietinum, C. montbretii) to over 5000 m (C. microphyllum) near glaciers in the Himalayas. The perennial species all die above ground after seed dispersal at the end of the season, and send out new stems the next season. Seeds are scattered in most cases, but rarely by more than a few meters. If ingested by animals, the seeds probably do not leave the alimentary canal undamaged, but hard seed coats may be more impenetrable than we think. Nobody has researched this aspect of dispersal. At any rate, dispersal is not very fast in nature. The distribution of Cicer spp. is given in Table 7 and Figure 8. The chickpea is grown outside the areas marked in Africa and the Americas, as man creates habitats where the crop thrives. Detailed locations are listed in van der Maesen (1972).

Many habitats remain virtually as they were many years ago, but populations of Cicer are always disjunct, and can be easily eliminated by overgrazing, landslides, and similar hazards. Except for a few species, none is really common, and grazing makes it difficult to collect or even to find undamaged specimens and seeds. C. subaphyllum Boiss. from the Kuh Ayub mountain near Persepolis, Iran, was found only once, in 1841, and may have been a single mutant form, and has since disappeared. There are several such rare species or very local endemics among the chickpea's relatives, and their existence is threatened because their maintenance is very difficult outside the ecological niche where they survive. Political accessibility is another problem; it is not confined to Cicer and is particularly acute at the time of writing.

Evolution and domestication

Recently, data have been gathered to improve our understanding of how the chickpea may have evolved from, or with, its nearest relatives. Several close relatives occur in SE Turkey and adjacent areas. Anatolia may have been the area from which chickpeas originated, where hunter-gatherers knew the progenitors in the wild and used them and took them into cultivation. Cicer reticulatum, the only species which so far has produced fertile offspring when crossed with chickpeas, seems a likely progenitor (Ladizinsky and Adler, 1975) and is classified in the primary gene pool (Ladizinsky, 1976 b). One can never be sure that it is the progenitor, since over 7000 years have elapsed since chickpeas were first cultivated. Archaeological finds of chickpeas are not as abundant as those of lentils, since the chickpea seed suffers more in a

Table 7. Distribution of Cicer spp.

	Afghanistan	Bulgaria	Egypt	Ethiopia	Greece	India	Iran	Iraq	Italy	Lebanon	Morocco	Pakistan	Spain	Sudan	Syria	Turkey	USSR C Asia	USSR Caucasia
C. acanthophyllum	*											*					*	
C. anatolicum							*	*								*		*
C. arietinum	*	*	*	*	*	*	*	*	*	*	*	*	*	*	*	*	*	*
C. atlanticum											*							
C. balcaricum																		*
C. baldshuanicum																	*	
C. bijugum								*							*	*		
C. chorassanicum	*						*											
C. cuneatum			*	*										*				
C. echinospermum																*		
C. fedtschenkoi	*																*	
C. flexuosum																	*	
C. floribundum																*		
C. garanicum																	*	
C. graecum					*													
C. grande																	*	
C. heterophyllum																*		
C. incanum																	*	
C. incisum						*		*								*		*
C. isauricum																*		
C. judaicum										*								
C. kermanense							*											
C. korshinskyi																	*	
C. macracanthum	*											*					*	
C. microphyllum	*					*						*					*	
C. mogoltavicum																	*	
C. montbretii		*		*												*		
C. multijugum	*															*		
C. nuristanicum	*					*						*						
C. oxyodon	*						*	*								?		
C. paucijugum																	*	
C. pinnatifidum								*							*	*		
C. pungens	*																*	
C. rechingeri	*																	
C. reticulatum																*		
C. songaricum																	*	
C. spiroceras						*												
C. stapfianum						*												
C. subaphyllum						*												
C. tragacanthoides						*											*	
C. yamashitae	*																	

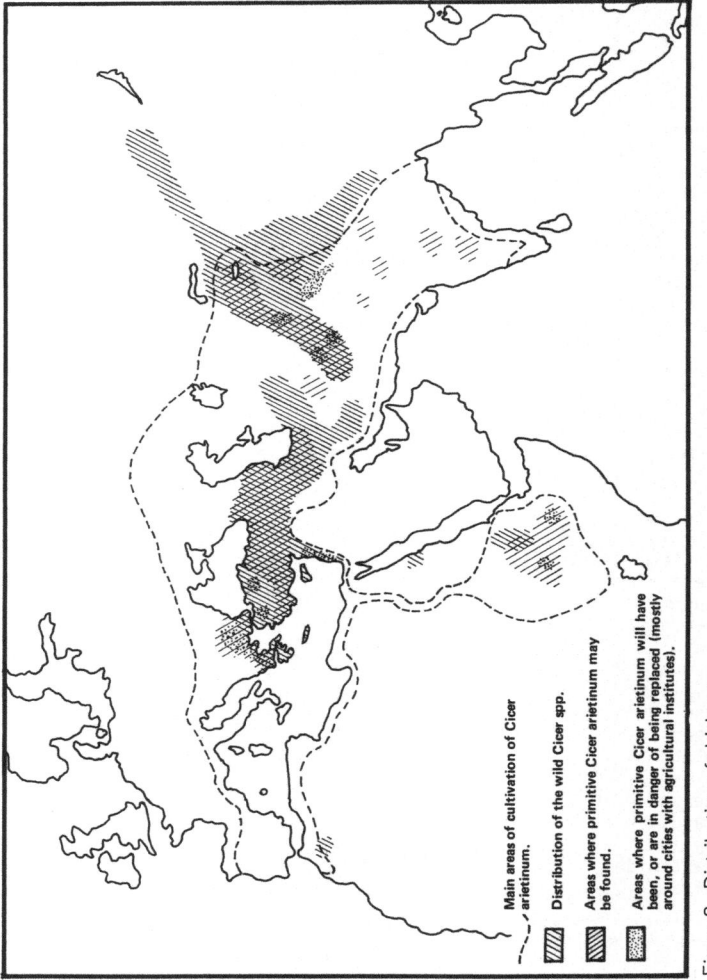

Figure 8. Distribution of chickpeas.

carbonised state; the beak often breaks off and the seed is then difficult to distinguish from a pea. The oldest prehistoric record of carbonised seeds, amassed in a human habitat, is from Haçilar, Turkey, dated at 5450 BC. In Egypt, there is a find of chickpeas from Deir-el-Medineh dating to 1400 BC. In Mesopotamia and Jericho, also, early finds were recorded. In Egypt, old papyri give the first written record of the crop (1580-1100 BC).

Cicer bijugum and C. echinospermum are morphologically as close to chickpea as C. reticulatum, but are not as close karyologically and chemically (Ladizinsky and Adler, 1975). Cicer echinospermum differs from the chickpea by a reciprocal translocation, and their hybrid was found to be sterile (Ladizinsky, 1976 b). C. echinospermum can also be placed in the secondary gene pool. All other species most probably belong to the tertiary gene pool.

The domestication of chickpea must have been accompanied by progress from shattering pods and prostrate stems to non-shattering pods and semi-erect to erect stems. In fact, C. reticulatum, C. bijugum, and C. echinospermum start off prostrate early in their life cycle, then send out their branches, and produce flowers and pods higher in the canopy. Very few chickpea cultivars are prostrate and these are obviously not preferred by cultivators. If left alone, they probably survive better than other cultivars, but they are no longer truly wild. The domestication of chickpea has also led to an increase in seed size.

Vavilov (1951) recognised five centres of origin for cultivated chickpea:

i. The Mediterranean (for large-seeded light-coloured forms).

ii. Central Asia.

iii. The Near East Centre (for pea-shaped forms - a secondary centre).

iv. The Indian (Hindustan) Centre.

v. A secondary centre in Ethiopia.

Rather than centres of origin, we acknowledge these areas now as centres of diversity. The core of the centres, southeast Turkey, is probably the earliest and original one, and it is in the Near East Centre that was described by Harlan (1971).

Diversity of the chickpea

A wide diversity exists in the cultivated chickpea, greater than in the wild species, in which only seed colour, size, and perhaps anthocyanin content distinguish accessions. Popova (1937) described 4 subspecies, 13 geographical proles, and 64 varieties in chickpea. The latter can better be considered as subraces because they differ in little more than seed and flower colour. The proles differ enough to be considered as races. Another overview of diversity in genetic resources is given by van der Maesen (1973 a), and diversity for breeding programs by Singh and Malhotra (Chapters 10 and 11).

Some observations on pests and diseases of wild chickpeas are given by van der Maesen (1980).

The evaluation of germplasm and breeding lines in areas where cultivars are needed is the first step in developing higher-yielding, better-adapted, and improved-quality lines. Local landraces are often yet unsurpassed in performance following thousands of generations of natural and unconscious selection!

References

Contandriopoulos, J., Pamukcuoglu, A. and Quezel, P. 1972. A propos des *Cicer* vivaces du pourtour mediterraneen oriental. Biologic Gallo-Hellenica 4: 3-18.

Harlan, J.R. 1971. Agricultural origins: centers and noncenters. Science 174: 468-474.

Kupicha, F. 1977. The delimitation of the tribe *Vicieae* and the relationships of *Cicer* L. Botanical Journal of the Linnaean Society 74: 131-162.

Ladizinsky, G. 1975. A new *Cicer* from Turkey. Notes of the Royal Botanical Garden, Edinburgh 34: 201-202.

Ladizinsky, G. 1976 a. The origin of chickpea, *Cicer arietinum* L. Euphytica 25: 211-217.

Ladizinsky, G. 1976 b. Genetic relationships among the annual species of *Cicer* L. Theoretical and Applied Genetics 48: 197-203.

Ladizinsky, G. and Adler, A. 1975. Seed protein electrophoresis of the annual species of *Cicer* L. Israel Journal of Botany 24: 183.

Popova, G.M. 1937. Chickpea. Pages 25-71 *in* Kulturnaya Flora SSR (Flora of cultivated plants of the USSR). Vol. 4, Grain Leguminosae, ed. Wulff, Z.V: Moscow and Leningrad State Agricultural Publishing Company.

van der Maesen, L.J.G. 1972. *Cicer* L., a monograph of the genus, with special reference to the chickpea (*Cicer arietinum* L.), its ecology and cultivation. Communication of the Agricultural University, Wageningen 72 - 10.

van der Maesen, L.J.G. 1973 a. Genetic resources of chickpea. Plant Genetic Resources Newsletter 30: 17-24.

van der Maesen, L.J.G. 1973 b. Chickpea: distribution of variability. Pages 30-34 in Survey of Crop Genetic Resources in their Centres of Diversity, ed. O.H. Frankel, Rome : FAO/IBP.

van der Maesen, L.J.G. 1980. Observations on pests and diseases of wild Cicer species. Indian Journal of Plant Protection 7: 39-42.

Vavilov, N.I. 1951. The origin, variation, immunity and breeding of cultivated plants. Chronica Botanica 13 (1/6): 14-54.

Appendix 9. The taxonomy of the genus <u>Cicer</u>.

Subgenus <u>Pseudononis</u> M.Pop.
Flowers small, 5-10(-15)mm, calyx subregular, its base hardly gibbous, teeth nearly equal, sublinear.

Section <u>Monocicer</u> M.Pop. Annuals, stems firm, erect or horizontal, branched from the base or at middle.

Series <u>Arietina</u> Linczevski, Lvs imparipinnate, arista small or absent:

<u>Cicer</u> <u>arietinum</u> L.

<u>Cicer</u> <u>bijugum</u> K.H.Rech.

<u>Cicer</u> <u>echinospermum</u> P.H.Davis.

<u>Cicer</u> <u>judaicum</u> Boiss.

<u>Cicer</u> <u>reticulatum</u> G.Ladizinsky

<u>Cicer</u> <u>pinnatifidum</u> Jaub. & Spach

Series <u>Cirrhifera</u> vdMaesen, Lvs ending in a tendril, arista short, 4-12 mm:

<u>Cicer</u> <u>cuneatum</u> Hochst.ex Rich.

Series <u>Macro-aristae</u> vdMaesen, Lvs imparipinnate, arista long, 5-20 mm:

<u>Cicer</u> <u>yamashitae</u> Kitam.

Section <u>Chamaecicer</u> M.Pop. Annuals or perennials, stem thin, creeping, branched, with a leafless part in the soil. Fls small, 5-10 mm.

Series <u>Annua</u> vdMaesen, annual:

<u>Cicer</u> <u>chorassanicum</u> (Bunge) M.Pop.

Series <u>Perennia</u> Lincz., perennial:

<u>Cicer</u> <u>incisum</u> (Willd.) K.Maly

Subgenus <u>Viciastrum</u> M.Pop.
Flowers medium large, 12-15 mm or large, 17-27 mm, calyx strongly gibbous at the base, teeth unequal. Perennials.

Section <u>Polycicer</u> M.Pop. Leaf rachis ending in a tendril or a leaflet, never a spine.

Subsection <u>Nano-polycicer</u> M.Pop. Rhizome creeping, stems short. Lvs imparipinnate, arista short, weak:

<u>Cicer</u> <u>atlanticum</u> Coss. ex Maire

Subsection <u>Macro-polycicer</u> M.Pop. Rhizome short, not creeping, stems ascending to 75 cm, arista firm, longer than pedicel.

Series <u>Persica</u> M.Pop. Inflorescences 1-2-flowered, fls 14-15 mm, calyx teeth 2-4 times the tube.   Stipules 14-15 mm, calyx teeth 2-4 times the tube, stipules c. half as large as the leaflets, which are in 2-12 pairs:

<u>Cicer</u> <u>kermanense</u> Bornm.

<u>Cicer</u> <u>oxyodon</u> Boiss. et Hohen.

<u>Cicer</u> <u>spiroceras</u> Jaub. et Spach

<u>Cicer</u> <u>subaphyllum</u> Boiss.

Series <u>Anatolo-Persica</u> (M.Pop.) Lincz. Inflorescences 1-2-flowered, fls large, 20-27 mm, calyx teeth short, stipules smaller than the largest leaflets, which are in 4-9 pairs:

<u>Cicer</u> <u>anatolicum</u> Alef.

<u>Cicer</u> <u>balcaricum</u> Galushko

Series <u>Europaeo-Anatolica</u> M.Pop. Inflorescences 2-5-flowered, bracts foliolate, stipules small or up to half as large as the leaflets, which are in 4-8 pairs:

<u>Cicer</u> <u>floribundum</u> Fenzl

<u>Cicer</u> <u>graecum</u> Orph.

<u>Cicer</u> <u>heterophyllum</u>

<u>Cicer</u> <u>isauricum</u> P.H.Davis

<u>Cicer</u> <u>montbretii</u> Jaub. et Spach

Series <u>Flexuosa</u> Lincz. Inflorescences 1-2-flowered, bracts minute, stipules much smaller than the leaflets, which are in 4-13 pairs:

<u>Cicer</u> <u>baldshuanicum</u> (M.Pop.) Lincz.

<u>Cicer</u> <u>flexuosum</u> Lipsky

<u>Cicer</u> <u>grande</u> (M.Pop.) Korotk.

<u>Cicer</u> <u>incanum</u> Korotk.?

<u>Cicer</u> <u>korshinskyi</u> Lincz.

<u>Cicer</u> <u>mogoltavicum</u> (M.Pop.) Koroleva

<u>Cicer</u> <u>nuristanicum</u> Kitamura

Series <u>Songorica</u> Lincz. Inflorescences 1-2-flowered, bracts minute. Stipules ± equal to the largest leaflets which are in 2-18 pairs:

<u>Cicer</u> <u>fedtschenkoi</u> Lincz.

<u>Cicer</u> <u>multijugum</u> van der Maesen

<u>Cicer</u> <u>paucijugum</u> Nevski

<u>Cicer</u> <u>songaricum</u> Steph. ex DC.

Series <u>Microphylla</u> Lincz. Inflorescences 1-2-flowered, bracts minute. Stipules smaller than the largest leaflets, which are in 7-10 pairs, or equal in size:

<u>Cicer</u> <u>microphyllum</u> Benth.

Section <u>Acanthocicer</u> M.Pop. Leaf rachis persistent, straight, ending in a
spine. Arista in a spine.   Stems branched, base woody, spiny perennial
shrublets. Series <u>Pungentia</u> Lincz. Stipules foliolate or small, spiny:

Cicer <u>pungens</u> Boiss.

Cicer <u>rechingeri</u> Podlech

Cicer <u>stapfianum</u> K.H.Rech.

Series <u>Macracantha</u> Lincz. Stipules spiny, quite long:

Cicer <u>macracanthum</u> M.Pop.

Cicer <u>acanthophyllum</u> Boriss.

Cicer <u>garanicum</u> Boriss.?

Series <u>Tragacanthoidea</u> Lincz. Stipule small, triangular incised perules:

Cicer <u>tragacanthoides</u> Jaub. et Spach.

10. COLLECTION AND EVALUATION OF CHICKPEA GENETIC RESOURCES

K.B. Singh

Plant Breeder, Chickpea (ICRISAT), ICARDA, Aleppo, Syria

R.S. Malhotra

Food Legume Improvement Program, ICARDA, Aleppo, Syria

Introduction

Chickpea (Cicer arietinum L.) ranks fifteenth among the food crops, and third among the pulse crops in the world. It ranks first among the food legumes grown in the region served by ICARDA, which extends from Morocco in the west to Pakistan in the east, and from Turkey in the north to Sudan in the south. There are two types of chickpea, desi (small seeded, angular and coloured) and kabuli (large seeded, ram-shaped and beige coloured). No precise information is available about the division of the 11 million hectares under chickpea between desis and kabulis, but an approximation is that more than 85 per cent of the area is covered with the desi type and the rest with kabuli. The desi type is generally grown in the Indian sub-continent and in Ethiopia, whereas the kabuli type is mainly cultivated in the Mediterranean region and Latin America. Since ICRISAT is located in India it has the sole responsibility for the improvement of the desi type. The joint responsibility for the kabuli type is shared between ICARDA and ICRISAT, and the work is based at ICARDA. Therefore, ICARDA serves as a world centre for the collection, maintenance and distribution of kabuli chickpea germplasm.

The origin of chickpea is believed to be in the region adjoining Turkey, Iran, Afghanistan and the USSR (van der Maesen, 1972; Ladizinsky and Adler 1976), and the greatest amount of genetic diversity should be found in this region. Since 85 per cent of world production comes from the Indian sub-continent, and most of the area is still sown with landraces, wide genetic diversity is also expected there. Besides these two major areas of genetic diversity, collecting in other west Asian countries, north Africa, south Europe and central and south America, where chickpeas have been grown for many years, may prove profitable.

Fortunately, much of the genetic diversity is still preserved, but there is a danger of genetic erosion, because plant breeding and the production of new cultivars has started in many countries. Therefore, there is a dire need for ICARDA, ICRISAT, IBPGR and national programs to make a concerted effort to collect as much of the genetic diversity as possible before it is lost for ever. Both the cultivated and the wild species should be collected.

J.R. Witcombe and W. Erskine (eds.) Genetic Resources and Their Exploitation - Chickpeas, Faba beans and Lentils. ISBN-13:978-94-009-6133-3 (PB)
©1984, Martinus Nijhoff/Dr W. Junk Publishers for ICARDA and IBPGR.

Table 8. Number of accessions in ICARDA chickpea germplasm collection by countries of origin.

Country	Number of accessions	Country	Number of accessions
AFRICA		NORTH & CENTRAL AMERICA	
Algeria	19	Ecuador	2
Egypt	50	Mexico	55
Ethiopia	29	United States of America	44
Malawi	1	SOUTH AMERICA	
Morocco	89	Chile	337
Sudan	7	Columbia	2
Tunisia	245	Peru	3
ASIA		EUROPE	
Afghanistan	868	Bulgaria	14
Cyprus	7	Czechoslovakia	6
India	243	France	3
Iran	1232	Greece	12
Iraq	30	Italy	19
Jordan	127	Portugal	5
Lebanon	26	Spain	213
Nepal	2	USSR	51
Pakistan	24	Yugoslavia	2
Syria	83		
ICARDA-Syria	1	Unknown	183
Turkey	386		
		TOTAL	4420

Collection

The intensive cultivation of only a few genotypes of a crop species and the adoption of a monoculture system increase the chance of serious crop losses from outbreaks of insect-pests and diseases. A varietal improvement program with a wide genetic base can help in preventing the rapid and extensive spread of insect-pests and diseases. To cope with the growing needs of varietal improvement programs, and to ensure their success, an assembly of the entire array of existing germplasm is essential.

The building of a germplasm collection can be achieved by various means, including exploration, exchange, and donation from international and national genetic resources units. Scientists working in national programs can obtain germplasm of both desi and kabuli types from ICRISAT, and kabuli germplasm from ICARDA. In addition, a large number of collections are available from national genetic resources units, including WRPIS, Washington, USA; the National Bureau of Plant Genetic Resources, New Delhi, India; the Aegean Regional Agricultural Research Institute, Izmir, Turkey; the College of Agriculture, Teheran University, Karaj, Iran and the N.I. Vavilov All Union Institute of Plant Industry, Leningrad, USSR.

ICARDA has assembled 4420 accessions of kabuli germplasm from 34 countries (Table 8). It seems that there is an adequate number of collections from most of the countries. There are, however, some obvious gaps for countries in Latin America and southern Europe. Future efforts should concentrate more on these areas. Most of these accessions have been assembled through personal contacts, seed exchange, and as the result of written requests. Only a small proportion of them have actually been collected in the field by ICARDA and ICRISAT. About 35 per cent of the accessions in the collection have been obtained from ICRISAT. In exchange, duplicates of most of the ICARDA accessions have been sent to ICRISAT. There is also an agreement to periodically exchange new kabuli accessions between the two centres.

ICARDA inherited, from the ALAD program in Lebanon, 30 samples of seven wild Cicer species: C. pinnatifidum, C. montbretii, C. judaicum, C. yamashitae, C. bijugum, C. cuneatum, and C. reticulatum. The evaluation of these for response to Ascochyta rabiei identified samples from C. pinnatifidum, C. montbretii, and C. judaicum as highly resistant. The reaction of accessions from the same species differed, which emphasises the need for larger collections of wild species and their evaluation for resistance and other traits.

The information on donor, pedigree, country of origin, etc., are entered in the accession record book. Such information is essential in cataloguing the germplasm accessions; sample pages from the Kabuli Chickpea Catalog (Singh et al., 1983) are shown in Tables 9 and 10.

Table 9. Sample page of chickpea passport information, donor and origin, for the ICARDA chickpea collection.

The following abbreviations are used in the headings:

ILC	ILC accession number
Don Org	Donor organisation
Col Org	Collecting organisation
Coll No	Collection number
CDat	Collection date
Cou	Country of origin
RDat	Date received by ICARDA/ALAD

ILC	Don Org	Don No	Col Org	Collector	Coll No (Pedigree)	CDat	Cou	Town/Province	RDat
1856	ICRISAT	ICC 2955	RPIP(D)	-	P-3387	-	IRN	-	0778
1857	ICRISAT	ICC 2966	MAI	-	P-3425(1561-1598)	-	IRN	MAMAGHAN	0778
1858	ICRISAT	ICC 2997	RPIP(D)	-	P-3467	-	IRN	-	0778
1859	ICRISAT	ICC 3042	RPIP(D)	-	P-3547	-	IRN	-	0778
1860	ICRISAT	ICC 3097	MAI	-	P-3610-1	-	IRN	KARAJ	0778
1861	ICRISAT	ICC 3243	MAI	-	P-3818(1944-1991)	-	IRN	JIROFT	0778
1862	ICRISAT	ICC 3268	MAI	-	P-3852(1979-2025)	-	IRN	TORBAT SHADMEHRE	0778
1863	ICRISAT	ICC 3304	MAI	-	P-3935(2064-2108)	-	IRN	TORBAT SHADMEHRE	0778
1864	ICRISAT	ICC 3315	MAI	-	P-3956(2084-2129)	-	IRN	TORBAT SHADMEHRE	0778
1865	ICRISAT	ICC 3344	-	-	P-4008(2132-2181)	-	IRN	-	0778
1866	ICRISAT	ICC 3382	-	-	P-4065-1	-	IRN	-	0778
1867	ICRISAT	ICC 4254	MAI	-	P-5163(2874-0929)	-	IRN	AHAR	0778
1868	ICRISAT	ICC 4258	MAI	-	P-5188(2986-0954)	-	IRN	AHAR	0778
1869	ICRISAT	ICC 4457	MAI	-	P-5467(309M)	-	IRN	NEYSHABUR	0778
1870	ICRISAT	ICC 4754	RPIP(K)	-	P-6366(0822-5057)	-	IRN	NEYSHABUR	0778
1871	ICRISAT	ICC 4928	PAU	-	C 104	-	IND	PUNJAB	0778
1872	ICRISAT	ICC 5247	PAU	-	GL 625	-	IND	PUNJAB	0778
1873	ICRISAT	ICC 5253	PAU	-	GL 633	-	IND	PUNJAB	0778
1874	ICRISAT	ICC 5298	PAU	-	L 287	-	IND	PUNJAB	0778
1875	ICRISAT	ICC 5299	PAU	-	L 299	-	IND	PUNJAB	0778
1876	ICRISAT	ICC 5336	PAU	-	NO.99	-	IND	MAHARASHTRA	0778
1877	ICRISAT	ICC 5339	PAU	-	NO.163	-	IND	MAHARASHTRA	0778
1878	ICRISAT	ICC 5341	PAU	-	NO.477	-	IND	MAHARASHTRA	0778
1879	ICRISAT	ICC 5407	FAU	-	PB-1	-	IND	LUDHIANA;PUNJAB	0778
1880	ICRISAT	ICC 5809	ARS	-	H.SAGAR-29-11	-	IND	MAHARASHTRA	0778
1881	ICRISAT	ICC 5842	ARS	-	NAMDURBAR-895	-	IND	MAHARASHTRA	0778
1882	ICRISAT	ICC 6011	ARS	-	WFWG 2	-	IND	MAHARASHTRA	0778
1883	ICRISAT	ICC 6012	ARS	-	WFWG 3	-	IND	MAHARASHTRA	0778
1884	ICRISAT	ICC 6041	JNKVV	-	JG 7	-	IND	MADHYA PRADESH	0778
1885	ICRISAT	ICC 6043	JNKVV	-	JG 9	-	IND	MADHYA PRADESH	0778
1886	ICRISAT	ICC 6051	JNKVV	-	JG 17	-	IND	MADHYA PRADESH	0778
1887	ICRISAT	ICC 6053	JNKVV	-	JG 20	-	IND	MADHYA PRADESH	0778
1888	-		-	-	ENTRY 2	-	ECU	-	0778
1889	ICRISAT	-	ICRISAT	LJGM	JM 2755-1	-	TUR	-	0778
1890	ICRISAT	-	ICRISAT	LJGM	JM 2755-2	-	TUR	-	0778

Table 10. Sample page of chickpea passport information, synonyms, for the ICARDA
          chickpea collection.

The following abbreviations are used in the headings:

NEC	The Near East Collection number.
PI	The PI number
Syn 3	Synonym 3. The Tal Amara (TA), Kfardan (KF) and Muslimiya (MS) numbers along with the year when the collection was grown at these locations in Lebanon and Syria.
Syn 4	Synonym 4. Any other number.

ILC	NEC	PI		SYN 3		SYN 4
386	1198	PI	360392	74TA	1810	-
387	1199	PI	360393	74TA	1812	-
388	1223	PI	360417	74TA	1864	-
389	1224	PI	360418	74TA	1865	-
390	1323	PI	360517	74TA	2062	-
391	1332	-		75TA	16621	-
392	1399	PI	360594	74TA	2159	-
393	1402	PI	360597	74TA	2160	-
394	1403	PI	360598	77MS	78619-1	-
395	1407	PI	360602	74TA	2166	-
396	1408	PI	360603	74TA	2167	-
397	1409	-		-		-
398	1410	PI	360605	74TA	2168	-
399	1415-1	PI	360610	74TA	2170	-
400	1415-2	PI	360610	74TA	2171	-
401	1417	PI	360612	77MS	78620-1	-
402	1419	PI	360614	74TA	2175	-
403	1422	PI	360617	74TA	2178	-
404	1424	PI	360619	74TA	2179	-
405	1426	PI	360621	74TA	2181	-
406	1428-1	PI	360623	74TA	2183	-
407	1428-2	PI	360623	74TA	2184	-
408	1431	PI	360626	74TA	2189	-
409	1454	PI	360649	74TA	2218	-
410	1460	PI	360655	74TA	2227	-
411	1461	PI	360656	77MS	78621-1	-
412	1462	PI	360658	74TA	2228	-
413	1464	PI	360660	74TA	2232	-
414	1467-1	PI	360689	74TA	2234	-
415	1467-2	PI	360689	77MS	78622-1	-
416	1468-1	PI	360690	77MS	78623-1	-
417	1468-2	PI	360690	74TA	2239	-
418	1472	PI	360702	77MS	78624-1	-
419	1474	-		77MS	78625-1	-
420	1475	-		77MS	78626-1	-

ILC	NEC	PI		SYN 3		SYN 4
351	1078	PI	360271	74TA	1598	-
352	1079	PI	360272	74TA	1602	-
353	1082	PI	360275	-		-
354	1085	PI	360278	77MS	78614-1	-
255	1086	PI	360279	74TA	1610	-
356	1090	PI	360283	-		-
357	1091	PI	360284	74TA	1623	-
358	1102	PI	360296	74TA	1638	-
359	1105	PI	360299	74TA	1644	-
360	1108	PI	360302	74TA	1646	-
361	1109	PI	360303	74TA	1648	-
362	1112-1	PI	360306	74TA	1654	-
363	1112-2	PI	360306	74TA	1656	-
364	1113	PI	360307	74TA	1659	-
365	1117	PI	360311	-		-
366	1119	PI	360313	74TA	1666	-
367	1121	PI	360315	74TA	1668	-
368	1124-1	PI	360318	74TA	1669	-
369	1124-2	PI	360318	74TA	1670	-
370	1124-3	PI	360318	74TA	1671	-
371	1126	PI	360320	77MS	78615-1	-
372	1129	PI	360323	77MS	78616-1	-
373	1131-1	PI	360325	74TA	1684	-
374	1131-2	PI	360325	74TA	1685	-
375	1133-1	PI	360327	74TA	1687	-
376	1133-2	PI	360327	74TA	1686	-
377	1134	PI	360328	74TA	1688	-
378	1139	PI	360333	77MS	78617-1	-
379	1141	PI	360335	74TA	1701	-
380	1142-1	PI	360336	74TA	1702	-
381	1142-2	PI	360336	74TA	1704	-
382	1151	PI	360345	74TA	1720	-
383	1157	PI	360351	74TA	1733	-
384	1162	PI	360356	74TA	1738	-
385	1197	PI	360391	74TA	1809	-

ILC	NEC	PI		SYN 3		SYN 4
316	1034-1	PI	360227	74TA	1513	-
317	1034-2	PI	360227	74TA	1514	-
318	1035	PI	360228	74TA	1516	-
319	1037	PI	360230	74TA	1517	-
320	1038	PI	360231	74TA	1518	-
321	1039	PI	360232	74TA	1520	-
322	1040	PI	360233	74TA	1521	-
323	1041	PI	360234	74TA	1523	-
324	1042	PI	360235	74TA	1524	-
325	1044	PI	360237	74TA	1526	-
326	1045	PI	360238	77MS	78612-1	-
327	1046-1	PI	360239	74TA	1531	-
328	1046-2	PI	360239	74TA	1532	-
329	1047	PI	360240	74TA	1534	-
330	1048-1	PI	360241	74TA	1535	-
331	1048-2	PI	360241	74TA	1537	-
332	1048-3	PI	360241	74TA	1538	-
333	1049	PI	360242	74TA	1540	-
334	1050	PI	360243	74TA	1542	-
335	1051	PI	360244	74TA	1544	-
336	1053	PI	360246	-		-
337	1055	PI	360248	74TA	1558	-
338	1056-1	PI	360249	74TA	1559	-
339	1056-2	PI	360249	74TA	1560	-
340	1057	PI	360250	74TA	1561	-
341	1059	PI	360252	74TA	1565	-
342	1060	PI	360253	74TA	1566	-
343	1061	PI	360254	74TA	1567	-
344	1062	PI	360255	74TA	1569	-
345	1063	PI	360256	74TA	1571	-
346	1065	PI	360258	-		-
347	1067	PI	360260	74TA	1578	-
348	1070	PI	360263	77MS	78613-1	-
349	1072	PI	360265	74TA	1587	-
350	1077	PI	360270	74TA	1597	-

Table 11. Ranges among kabuli chickpea accessions in the ICARDA collection for 19 characters.

Character	Range
Time to flower (days)	70-94
Flower duration (days)	11-36
Time to maturity (days)	114-124
Plant height (cm)	15-50
Canopy width (cm)	15-60
Primary branches/plant	2.3-16.0
Secondary branches/plant	0.3-22.7
Tertiary branches/plant	0.0-7.3
Pods/plant	4-100
Seeds/pod	0.1-3.1
Biological yield (g/m^2)	35-533
Grain yield (g/m^2)	7-292
Harvest index (%)	7-76
100-seed weight (g)	8.7-59.1
Protein content (%)	16.0-24.8
Resistance to seed shattering	HR-S *
Tolerance to cold	T-HS *
Resistance to blight	R-HS *
Tolerance to iron deficiency	T-S *

* HR = Highly resistant,
  R = Resistant,
  T = Tolerant,
  S = Susceptible,
  HS = Highly susceptible.

Evaluation

Introduction

The evaluation of germplasm is the key to its utilization in breeding programs. It is desirable that the evaluation should be undertaken at many locations in different agro-climatic conditions, but this can rarely be done, mainly due to limited resources. Indeed, the evaluation often has to be done at a single location in several batches over a number of years. For more efficient evaluation, 'hot-spots' are used for screening for reaction to disease or insect-pests.

Evaluation is a continuous process, because the focus of breeding shifts with the passage of time and the exposure of the crop to changing environmental and agro-climatic conditions. For instance, <u>Botrytis cinerea</u>, grey mould of chickpea, was not a serious problem about a decade ago, but with the introduction of new cultivars it has become more of a problem. Possibly, the new cultivars are more susceptible to this pathogen. Similarly, chickpeas in certain areas are prone to the effects of low temperature, drought, leaf miners, pod borers, ascochyta blight, wilts, root rots, and other stress conditions. Evaluation should be problem-oriented, and a multi-disciplinary approach should be followed with, preferably, pathologists, entomologists and microbiologists cooperating with the breeders to screen the germplasm effectively.

The evaluation of chickpea germplasm includes the following group of characters: morphological traits, stress characters, yield characters, seed characters, and quality characters. Some of these are listed in Table 11 and in Chapter 11.

After evaluation, the characterization of genetic stocks into different groups will help in the efficient testing and utilization of material, without wasting resources to evaluate the entire gene pool. The characterization thus helps in the testing of only small, desirable lots of material, on a large scale.

Evaluation at ICARDA

Over 3300 kabuli chickpea accessions, maintained at ICARDA, have been evaluated at Tel Hadya, Syria, for 29 characters in the past four seasons. Wide genetic variation has been observed for all of them. The range for 19 characters is shown in Table 11 and an example page of the chickpea evaluation data is shown in Table 12.

Table 12. Sample of the chickpea evaluation data for the ICARDA chickpea collection.

The following abbreviations are used in the headings:

ILC	ILC number
DaF	Days to 50% flowering
FDu	Flowering duration
FCo	Flower colour
DMa	Days to maturity
Hgt	Plant height
CaW	Canopy width
PrB	Primary branches per plant
SeB	Secondary branches per plant
TrB	Tertiary branches per plant
GrH	Growth habit
LfS	Leaf size
P/P	Pods per plant
S/P	Seeds per pod
PDe	Pod dehiscence
BiYd	Biological yield
SeYd	Seed yield
HI	Harvest index
100W	100-seed weight
SSh	Seed shape
SCo	Seed colour
SRo	Roughness of seed surface
Pro %	Protein content of seed
CSL	Cold susceptibility at low altitude
CSM	Cold susceptibility at medium altitude
CSH	Cold susceptibility at high altitude
FSS	Frost susceptibility at seedling stage
FSP	Frost susceptibility at pre-flowering stage
FeD	Susceptibility to iron defficiency
AB1	Susceptibility to ascochyta blight

ILC	DaF	FDu	FCo	DMa	Hst	CaW	PrB	SeB	TrB	GrH	LfS	P/P	S/P	PDe	BiYd	SeYd	HI	100W	SSh	SCo	SRo	ProZ	CSL	CSM	CSH	FSS	FSP	FeD	Abl
1996	80	26	W	120	31	40	8.3	5.3	1.3	SS	M	23	0.8	1	1110	617	56	27.3	OWL	BE	R	19.1	1	3	9	3	7	1	9
1997	83	23	W	114	31	45	9.0	5.3	0.7	SS	M	28	1.1	1	410	250	61	19.6	OWL	BE	R	19.4	3	1	9	1	7	1	9
1998	78	28	W	114	30	45	5.0	3.3	0.3	S	M	30	1.0	1	1060	574	54	27.7	OWL	BE	R	20.6	3	1	9	2	7	1	9
1999	83	23	W	114	32	32	7.0	7.0	0.0	SS	M	17	1.2	1	840	389	47	19.9	OWL	BE	R	20.2	3	1	9	–	–	1	9
2000	80	21	W	114	27	42	5.3	5.7	0.0	SS	M	19	1.1	1	700	377	54	23.1	OWL	BE	R	18.8	1	3	9	–	7	1	9
2001	76	25	W	114	26	43	2.3	2.3	0.3	SS	M	21	1.2	1	900	500	56	18.7	OWL	BE	R	19.9	3	3	9	2	7	1	9
2002	83	23	W	120	30	42	7.7	5.0	0.0	SS	S	16	1.0	1	480	233	49	18.2	OWL	YE	R	20.6	1	1	7	1	5	1	9
2003	80	26	W	120	31	40	6.0	6.0	0.3	SS	M	35	0.9	1	620	338	55	23.5	OWL	BE	R	19.0	3	3	7	–	–	1	7
2004	76	30	W	120	29	45	6.3	5.3	0.0	SS	M	27	1.0	1	540	316	59	27.2	OWL	BE	R	18.6	3	1	7	–	7	1	9
2005	83	23	W	120	36	46	8.7	9.0	0.0	SS	M	25	0.8	1	550	282	51	17.2	OWL	YE	R	19.3	1	3	9	–	5	1	9
2006	83	18	W	120	24	45	6.0	4.3	0.0	SS	M	38	1.0	1	730	414	57	20.1	OWL	BE	R	19.5	1	3	9	–	–	1	9
2007	80	21	W	120	30	43	4.7	4.3	0.3	SS	S	25	1.0	1	740	402	54	22.4	OWL	BE	R	22.3	3	1	9	–	–	1	9
2008	83	18	W	120	34	41	5.7	3.0	0.0	SS	M	42	0.7	2	480	259	54	21.5	PEA	BE	S	19.6	1	1	9	–	6	1	9
2009	76	30	W	114	33	45	7.3	7.7	0.0	SS	M	36	0.7	2	860	464	54	20.0	OWL	BE	R	19.0	3	1	9	–	–	1	9
2010	76	25	W	120	25	45	3.3	2.7	0.0	SS	M	25	1.4	1	480	245	54	14.9	OWL	BE	R	22.6	3	1	9	–	6	1	9
2011	83	18	W	114	23	45	3.7	2.7	0.0	S	M	29	1.2	1	480	276	58	19.9	OWL	BE	R	18.8	3	1	9	–	6	1	9
2012	80	21	W	114	25	39	6.7	4.0	0.0	S	M	25	1.0	1	500	283	53	17.1	OWL	BE	R	19.5	1	1	9	–	6	1	9
2013	80	26	W	120	30	41	5.0	2.7	0.3	SS	M	24	1.0	1	606	302	50	26.4	OWL	YE	R	20.3	1	1	9	–	6	1	9
2014	80	21	W	114	26	40	3.0	1.7	0.3	SS	M	26	0.9	1	700	376	54	19.7	OWL	BE	R	20.5	3	1	9	–	–	1	9
2015	72	29	W	114	26	45	4.3	3.3	0.0	SS	M	51	1.1	1	600	324	54	16.3	OWL	BE	R	19.4	3	3	9	–	6	1	9
2016	76	26	W	114	26	45	4.0	4.0	0.3	SS	M	27	1.3	2	640	353	55	15.9	OWL	BE	R	20.2	2	1	9	–	7	1	9
2017	76	26	W	114	25	43	4.7	3.0	0.0	S	M	31	1.2	2	850	461	54	19.9	OWL	BE	R	19.3	3	1	9	–	–	1	9
2018	76	25	W	114	24	42	4.0	4.3	0.0	S	M	20	0.6	2	520	257	49	15.5	OWL	BE	R	20.6	3	3	9	–	7	1	9
2019	76	25	W	114	25	43	3.3	3.0	0.0	SS	M	32	1.2	2	640	326	51	18.7	OWL	BE	R	20.2	3	1	9	–	7	1	9
2020	76	28	W	114	26	42	3.7	1.7	0.0	SS	M	23	1.2	2	440	230	52	19.7	OWL	BE	R	19.7	1	1	9	–	6	1	9
2021	76	25	W	114	25	37	6.0	8.7	1.0	SS	M	54	1.0	2	533	291	55	20.7	OWL	BE	R	19.6	3	1	9	–	7	1	9
2022	76	25	W	114	23	41	5.7	4.3	0.7	SS	M	43	1.0	1	367	200	55	20.8	OWL	BE	R	20.5	1	3	9	–	6	1	9
2023	76	25	W	114	24	41	6.0	2.7	0.7	S	M	31	0.8	1	440	262	62	22.2	OWL	BE	R	21.4	1	3	9	–	7	1	9
2024	76	25	W	114	24	41	5.0	2.7	0.0	SS	M	36	1.1	1	700	320	46	20.6	OWL	BE	R	22.5	3	1	9	–	7	1	9
2025	76	25	W	114	25	32	5.0	5.0	0.0	SS	M	30	1.1	1	600	230	38	29.7	OWL	YE	R	23.7	1	–	9	–	7	1	7
2026	76	25	W	114	27	42	5.3	4.0	0.7	SS	M	13	1.3	1	500	264	53	19.7	OWL	BE	R	20.4	3	1	9	–	7	1	9
2027	80	26	W	114	27	40	5.0	4.3	0.3	S	M	39	1.0	1	630	322	51	19.6	OWL	BE	R	19.6	3	3	9	–	7	1	9
2028	80	21	W	114	27	45	5.0	3.0	0.0	S	M	30	1.3	1	1040	592	57	20.6	OWL	BE	R	20.1	3	3	9	–	6	1	9
2029	80	21	W	114	26	46	4.3	3.3	0.0	S	M	25	1.1	1	450	266	59	20.1	OWL	BE	R	19.2	3	1	9	–	6	1	9
2030	83	18	W	114	26	50	4.3	3.0	0.3	SS	M	36	1.0	1	780	423	54	20.1	OWL	BE	R	19.8	3	1	9	–	6	1	9

The evaluation of some of the traits is detailed below:

Ascochyta blight

Blight caused by Ascochyta rabiei is the major disease in west Asia, north Africa, south Europe, Pakistan and north-west India. The yield loss caused by this disease can reach up to 100 per cent during epiphytotic years. Therefore, great efforts are being devoted to resistance breeding. A reliable, large-scale, screening technique for the field has been developed at ICARDA. This technique involves spreading infected disease-debris between the accessions in the field, growing frequent spreader rows, providing sprinkler irrigation to create high humidity for disease development and spread, and, if required, spraying with a picnospore suspension (Singh et al., 1981).

Using this technique, about 14000 germplasm accessions of kabuli and desi types have been screened, and a number of resistant sources have been identified. The material was scored on a 1-9 scale where 1= no visible lesions on any plants (highly resistant), 3 = lesions visible on less than 10% of the plants, no stem girdling (resistant), 5 = lesions visible on up to 25% of the plants, stem girdling on less that 10% of the plants but little damage (tolerant), 7 = lesions on most plants, stem girdling on less than 50% of the plants resulting in the death of a few plants (susceptible), 9 = lesions on all plants, stem girdling on more than 50% of the plants and death of most plants (highly susceptible).

The evaluation of over 3800 germplasm accessions helped in the identification of a large number of lines resistant to this disease. Some of the better sources of resistance are ILC 72, ILC 191, ILC 196, ILC 200, ILC 201, ILC 202, ILC 2506, ILC 2956, ILC 3279, ILC 3346, ILC 3400 and Pch 128.

Wilt

This disease is caused by Fusarium oxysporum f. sp. ciceri and it has been reported from Bangladesh, Burma, Ethiopia, India, Malawi, Mexico, Pakistan, Peru, Sudan, Tunisia, and the USA (Nene, 1980). It causes substantial losses and can be devastating. The fungus is soil borne and can also spread through seed. Both field and laboratory screening techniques have been described by Nene et al. (1981), who have also found a number of sources of resistance.

Leaf miner

The leaf miner, Lyriomyza cicerina is reported to be confined to the Mediterranean region. The extent of damage is variable, but it does cause damage of up to 20 per cent in some years. Over 2500 germplasm accessions were evaluated during the 1981-82 season.

Pod borer

Pod borers (<u>Heliothis</u> spp.) have been reported from many countries but are particularly destructive in the tropics. A screening technique based on natural infestation was developed at ICRISAT.

Cold tolerance

The chickpea crop is traditionally sown during spring in the Mediterranean region, whereas research conducted during recent years at ICARDA indicates that the yield can be doubled if the sowing date is advanced to winter. The advance in sowing date is only possible with cultivars having resistance to ascochyta blight and tolerance to cold. Screening for cold tolerance in chickpeas was begun in 1978-79; the 1981-82 season was ideal for this work as sub-zero temperatures were recorded at Tel Hadya for 31 nights, as compared to less than 20 nights in any of the previous three seasons. The complete death of plants occurred during this year, while in previous years only partial damage to plants was recorded. Over 10000 accessions were evaluated on a 1-9 scale, where 1 = no visible symptom of damage on any plant, 3 = up to 10% leaflets in most plants show yellowing, but there is no damage to the stem; 5 = 10 to 25% leaflets show damage, and up to 10% of the plants with stem breakage; 7 = 25-75% leaflets and, stems damaged, and up to 50% plants killed above ground level but later most recovered; 9 = all leaflets and stems above ground level damaged, and all plants killed.

As well as at Tel Hadya, the evaluation of chickpeas for cold tolerance has been done at Terbol (Beqa'a valley of Lebanon) and Hymana (near Ankara in Turkey) in different years including 1978-79, 1979-80 and 1981-82. A large number of accessions have been found to be cold tolerant and some of the most tolerant are: ILC 410, ILC 2406, ILC 2479, ILC 2491, ILC 2529 and ILC 2636.

<u>Orobanche</u> spp.

<u>Orobanche</u> <u>crenata</u> is a serious parasite of many legumes in the Mediterranean region. This parasite rarely affects spring sown chickpeas, but when the crop is winter sown isolated plants are found to be infected. The reaction of germplasm to <u>Orobanche</u> is now being studied in a winter sown crop growing in heavily infested soil at Kafr Antoon, northern Syria.

Photoperiod

Photoperiod insensitivity is associated with the wide adaptation of genotypes. In chickpeas, it has an additional significance, most of the sources of resistance to ascochyta blight are highly sensitive to photoperiod when grown during the off-season, from June to October, in decreasing day length in west Asia. Until a more refined technique is developed, lines are being scored by growing them during the off-season. The highly sensitive lines do not mature, less sensitive lines attain physiological maturity, and the least sensitive lines mature fully before the onset of cold weather at the off-season sites, which are cooler than the normal chickpea growing areas.

The screening technique for this character is still under development. The work accomplished so far indicates that the early maturing lines originating from Sudan and India are least sensitive to photoperiod, and the late maturing lines originating from the USSR are the most sensitive.

Cooking quality

Hommos Bi-tehineh

Chickpeas are consumed most widely in parts of west Asia as 'Hommos Bi-tehineh'. In its preparation, one kilo of chickpeas are soaked overnight in 3 litres of water with the addition of 20 g sodium bicarbonate. The soaked chickpeas are boiled for 15 minutes in a pressure cooker, then cooled and strained. The cooked chickpeas, 300 g of tehineh (a preparation made from crushed sesame seeds), 20 g salt, 20 g citric acid, and 700 ml water are mixed for 5 minutes in a Wareing blender at room temperature. Organoleptic tests are then carried out to measure the acceptability of the samples (Tannous and Singh, 1980). Utilizing this technique a number of samples have been scored.

Hommos Bi-tehineh prepared from desi types was of poor quality and was often unacceptable. Surprisingly, this product prepared from decorticated desi types was evaluated as being as good as those prepared from kabuli types. Though Hommos Bi-tehineh prepared from intermediate-sized desi chickpeas was found acceptable, it was rated inferior to that of kabulis. There was no difference between the Hommos Bi-tehineh prepared from small and large seeded kabuli chickpeas.

Cooking time

A repeatable procedure for determining cooking time has been developed as follows: A small sample cup with a capacity of 25 ml is filled with chickpeas, 200 ml of tap water is added and the chickpeas are boiled under reflux on a 'Labconco' crude fibre apparatus, using 600 ml beakers. The samples are started at 2 minute intervals to facilitate continuous throughput. After 45 minutes, a chickpea seed from the first sample is withdrawn without interrupting the boiling. The degree of cooking is tested by visual determination of the degree of gelatinisation and the softness of the seeds, as determined by finger pressure. If the seed is not cooked, another seed is tested after a further 10 minutes. This procedure is continued until half the seeds that have been tested are recorded as cooked. At that stage the degree of cooking is verified by testing up to 5 seeds. If the consensus shows that the seeds are not cooked, further seeds are tested until at least 5 seeds indicate that cooking is complete, at which time total cooking time is recorded.

Sufficient variability for cooking time was found. In general, small seeded chickpeas took less time to cook than large seeded.

Plant height

Tall types: Some of the lines found to be tall were: ILC 72, ILC 196, ILC 197, ILC 198, ILC 199, ILC 201, ILC 202, ILC 779, ILC 2951, ILC 2952, ILC 2953, ILC 2955, ILC 2956, ILC 2957, ILC 3272, ILC 3273, ILC 3274, ILC 3279, and ILC 3346.

Seed weight

High 100-seed weight: The accessions with a 100-seed weight heavier than 55g were: ILC 95, ILC 96, ILC 97, ILC 99, ILC 100, ILC 101, ILC 148, ILC 149, ILC 445, ILC 470, ILC 1250, ILC 1253, ILC 1254, ILC 2398, and ILC 2593.

Data storage and retrieval

More use can be made of the evaluation data when they are stored in a computer so that useful information can be extracted with ease. The evaluation data collected at ICARDA have been stored in a VAX-11/780 computer and programs to select sub-sets of accessions having specific characters, or combinations of characters, are in use. This enables chickpea scientists to obtain the desired material for their programs. The passport information and evaluation data for 29 characters are published (Singh, Malhotra and Witcombe, 1983).

The evaluation data can also be analysed. For example, an analysis of variance by country of origin (Table 13) revealed clear geographic differences between the accessions.

Conclusions

It is estimated that 15 per cent of the world chickpea production, or a little over one million tonnes of seed, comes from kabuli chickpeas. The average productivity is estimated to be around 800 kg/ha.

About five years ago, approximately 3400 kabuli germplasm accessions were already in collections; half of these were maintained by ALAD and the others were with ICRISAT. All of these are now assembled at ICARDA, and a further 1000 accessions have been added to bring the total to 4420. Despite the large size of the collection, it is far from being an adequate sample of the variability and there are many gaps. Fortunately, natural variability is still preserved, because most of the area under chickpea cultivation is still sown to landraces.

Table 13. Means and standard deviations according to country of origin for some continuously varying characters.

	DaF* Mean	DaF* S.D.	FID Mean	FID S.D.	DaM Mean	DaM S.D.	Hgt Mean	Hgt S.D.	CaW Mean	CaW S.D.	PrB Mean	PrB S.D.	SeB Mean	SeB S.D.	P/P Mean	P/P S.D.	S/P Mean	S/P S.D.
AFG	80.1	0.1	22.4	0.2	115.8	0.1	29.4	0.1	38.9	0.2	6.2	0.1	5.1	0.1	**26.5**	0.3	1.27	0.01
CHL	81.6	0.5	20.6	1.0	119.6	0.4	31.1	0.6	**41.6**	0.9	**7.3**	0.5	**7.7**	1.0	23.1	2.3	1.19	0.05
DZA	82.2	0.8	21.8	0.9	120.0	**	32.3	0.6	37.6	1.2	5.9	0.4	4.6	0.5	19.3	1.6	0.94	0.03
EGY	77.2	0.3	**28.8**	0.4	119.4	0.3	28.6	0.5	31.9	0.8	4.9	0.2	2.9	0.1	21.2	0.9	**1.39**	0.05
ESP	78.8	1.2	21.6	0.9	120.3	0.7	31.4	0.8	38.5	1.2	5.8	0.5	4.9	0.5	14.9	1.6	1.16	0.10
ETH	80.4	0.5	23.2	0.5	117.0	0.6	30.7	0.7	40.6	1.6	6.6	0.3	5.5	0.8	**28.0**	1.5	1.24	0.06
GRC	80.4	1.1	20.9	1.2	117.6	1.0	32.0	0.8	41.4	1.3	5.9	0.4	5.1	0.7	18.3	2.4	1.07	0.09
IND	81.2	0.2	22.4	0.2	117.8	0.2	29.0	0.3	37.8	0.5	5.6	0.1	4.8	0.2	22.2	0.5	1.13	0.02
IRN	82.4	0.1	23.0	0.1	118.6	0.1	29.1	0.1	**41.7**	0.1	6.6	0.1	6.4	0.1	23.1	0.3	1.12	0.01
IRQ	80.1	0.6	21.9	0.7	120.0	**	27.1	0.4	38.2	0.8	**7.3**	0.5	6.3	0.6	16.6	1.1	0.93	0.04
JOR	82.1	0.4	19.6	0.4	119.4	0.3	27.8	0.6	41.0	0.8	5.9	0.2	4.6	0.4	20.0	1.3	0.94	0.04
LBN	81.8	0.7	23.3	0.7	119.8	0.2	29.0	0.6	39.8	0.8	6.4	0.3	6.0	0.6	21.2	1.2	1.04	0.05
MAR	80.6	0.7	22.7	0.6	118.1	0.6	29.9	0.5	39.5	1.0	6.2	0.3	5.9	0.8	22.0	1.2	1.11	0.10
MEX	73.9	1.6	23.6	1.2	117.0	1.1	30.0	0.4	39.6	1.2	6.3	0.6	4.8	0.2	21.3	3.7	**1.50**	0.26
PAK	**83.3**	0.4	21.7	0.6	118.0	0.6	27.4	0.7	32.5	1.1	5.3	0.5	4.9	0.7	20.3	1.8	1.15	0.07
SUN	**82.8**	0.5	21.7	0.4	119.0	0.4	**36.6**	1.1	39.3	1.1	6.1	0.3	4.6	0.5	20.5	0.9	1.14	0.05
SYR	80.5	0.5	21.2	0.4	119.2	0.2	28.7	0.3	41.4	0.7	6.6	0.1	6.2	0.4	18.9	0.7	0.98	0.03
TUN	80.6	0.5	21.6	0.6	**120.7**	0.3	31.5	0.5	40.5	0.7	5.8	0.2	5.2	0.4	16.1	0.9	1.08	0.03
TUR	80.3	0.2	22.4	0.2	118.9	0.1	30.6	0.2	40.7	0.3	5.7	0.1	5.0	0.1	20.2	0.4	1.07	0.01
USA	79.1	0.6	23.0	0.5	116.6	0.7	30.6	0.7	40.0	1.0	5.6	0.3	4.4	0.4	21.1	1.5	1.00	0.05
OVERALL	81.2	0.1	22.7	0.1	117.9	0.1	29.5	0.1	40.2	0.1	6.2	0.1	5.6	0.1	25.1	0.2	1.11	0.01
F	21.3		14.2		42.5		21.3		19.4		8.0		10.3		26.6		15.3	

* A key to the character abbreviations is given in Table 12.
** Indicates the character is invariate and truely zero.

There is an urgent need to assemble them before they are lost for ever. The obvious gaps are in areas of north Africa, south Europe, and Latin America. These areas should receive priority in collection of kabuli chickpeas.

Regarding the collection of wild _Cicer_ species, only a small beginning has been made. One reason seems to be the lack of enthusiasm on the part of plant breeders to use them, because of non-crossability of the cultivated species with the wild species; the only exception is in crosses between _C. arietinum_ and _C. reticulatum_. With recent advances in protoplast fusion it should be possible, in the future, to cross the cultivated species with several of the 38 wild species. Therefore, special attention needs to be given to their collection. The evaluation of collected wild species has been limited to wilt and ascochyta blight diseases but it is clear that evaluation of these species should be done for more characters of interest.

Although most of the accessions available at ICARDA have been evaluated for 29 characters, this has been done largely only at Tel Hadya. Only part of the collection has been evaluated for resistance to ascochyta blight and tolerance to cold at more than one location. The value of the collection will increase if the evaluation is done at more than one location.

Systematic evaluation of germplasm is lacking for many useful characters, such as photoperiod, nodulating capability, ability to flower and fruit during relatively low temperatures, and drought tolerance. These should receive attention in the future.

References

Ladizinsky, G. and Adler, A. 1976. The origin of chickpea _Cicer_ _arietinum_ L. Euphytica 25: 211 - 217.

Nene, Y.L. 1980. A world list of pigeonpea (_Cajanus_ _cajan_ (L.) Millsp.) and chickpea (_Cicer_ _arietinum_ L.) pathogens. Pulse Pathology Progress Report - 8. Hyderabad, India: ICRISAT.

Nene, Y.L., Haware, M.P. and Reddy, M.V. 1981. Chickpea Diseases. Resistance Screening Techniques. Information Bulletin No. 10. Hyderabad, India: ICRISAT.

Singh, K.B., Hawtin, G.C., Nene, Y.L. and Reddy, M.V. 1981. Resistance in chickpeas to _Ascochyta_ _rabiei_ (Pass.) Lab. Plant Disease 65: 586-587.

Singh, K.B., Malhotra, R.S. and Witcombe, J.R. 1983. Kabuli Chickpea Germplasm Catalog. Aleppo, Syria: ICARDA.

Tannous, R. and Singh, K.B. 1980. Organoleptic evaluation of 'Hommos Bi-tehineh' prepared from chickpeas. International Chickpea Newsletter 2: 20 - 21.

van der Maesen, L.J.G. 1972. _Cicer_ L. Monograph of the genus with special reference to the chickpea (_Cicer arietinum_ L.), its ecology, and cultivation. Wageningen, The Netherlands: Veenman and Zonen.

11. EXPLOITATION OF CHICKPEA GENETIC RESOURCES

K.B.Singh

Plant Breeder, Chickpea (ICRISAT), ICARDA, Aleppo, Syria

R.S. Malhotra

Food Legume Improvement Program, ICARDA, Aleppo, Syria

Introduction

During the 1960s and 1970s a concerted effort was made to assemble collections of diverse germplasm of chickpea. As a result, over 12,000 kabuli and desi germplasm accessions have been assembled at ICRISAT, and over 4,400 kabuli accessions at ICARDA, these kabuli accessions have been evaluated for 29 characters, mainly at Tel Hadya in northern Syria (Singh and Malhotra, Chapter 10). Desirable genetic variability has been found for almost all the characters for which the kabuli accessions have been evaluated. Thus the stage is set for the utilization of germplasm; the national breeding programs in the countries served by ICARDA can make better use of the assembled germplasm for the development of cultivars during the 1980s.

Naturally occurring genetic variability receives the first attention of any plant breeding program. This variability has been exploited at ICARDA, primarily in two ways. First, certain germplasm lines have been directly used for the development of cultivars for commercial production. This is the cheapest and quickest way of developing cultivars for direct use by the farmers. Second, genetic variability has been exploited in a hybridisation program to incorporate genes for characters, such as resistance, into high yielding, but susceptible, cultivars.

Three major aspects of work in genetic resources are collection, evaluation, and utilization (Figure 9). In Singh and Malhotra (Chapter 10) collection and evaluation are discussed.

Direct exploitation for cultivar development

One of the major uses of germplasm collections has been the development of cultivars for commercial cultivation. There are a large number of examples of the development of cultivars through the direct exploitation of introduced germplasm, and the steps involved in chickpea cultivar development will be briefly described here.

J.R. Witcombe and W. Erskine (eds.) Genetic Resources and Their Exploitation - Chickpeas, Faba beans and Lentils. ISBN-13:978-94-009-6133-3 (PB)
©1984, Martinus Nijhoff/Dr W. Junk Publishers for ICARDA and IBPGR.

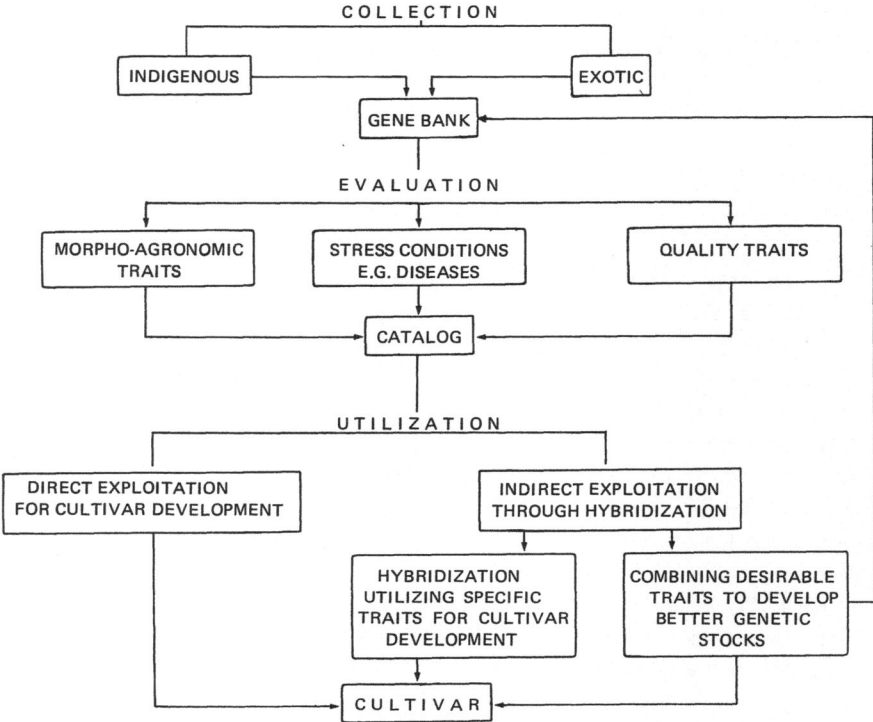

Figure 9.   Pathways for collection, evaluation and utilization of germplasm.

Step 1 - Single observation rows

The introduced germplasm is grown in single observation rows 2-4 m long. One or two check cultivars are sown at intervals of every 5 or 10 lines. If there are two check cultivars they are alternated in the planting scheme.

If disease resistance is a requirement in new cultivars, epiphytotic disease conditions are created. The normal cultural practices for the chickpea crop in the area are followed in respect of sowing date, fertilization and irrigation, and in disease, insect-pest and weed control.

A set of descriptors is recorded for each accession, and a few that have been used at ICARDA are listed below:

i. Date of emergence.

ii. Time to 50% flowering: the number of days from the date of sowing to the date on which 50% of the plants in an accession start blooming.

iii. Branching pattern: the number of primary, secondary, and tertiary branches are recorded on an average of 3-5 plants per accession.

iv. Growth habit: the chickpea plant is a bush type and semi-spreading, but occasionally there are erect, semi-erect, and spreading types.

v. Plant height: recorded from the ground level to the highest point in the plant without disturbance. The average of 3-5 plants is taken as the height of the accession.

vi. Screening for disease resistance: two diseases, ascochyta blight and wilt complex, seriously affect chickpea production. If either of these are important in an area, the lines should be screened by promoting the disease development under epiphytotic conditions. A 1-9 scale for scoring is used, with 9 the most susceptible.

vii. Screening for insect resistance: work on breeding for Heliothis spp. resistance is in progress at ICRISAT, and lines differing in susceptibility are available. Work on resistance to leaf miner (Lyriomyza cicerina) has been initiated at ICARDA. If either of these insect-pests are important in an area, the lines should be scored on a 1-9 scale, with 9 the most susceptible.

viii. Pod number: the number of pods per plant is recorded on an average of 3-5 plants.

ix. 100-seed weight: recorded in grams on a random sample of 100 seeds.

x. Any special character: any other characters that may be important in a given region, such as pod shedding, pod dehiscence, lodging score, cold tolerance, and tolerance to iron deficiency, may be recorded.

xi. Uniformity: the lines may be noted as uniform or non-uniform for different characters.

xii. Time to maturity: the number of days taken from the date of sowing to the date when 90% of the plants have matured.

xiii. Yield: judged by an "eye ball" evaluation on a 1-9 scale.

xiv. Seed quality: seed type, colour, shape and surface are determined in the laboratory.

xv. Remarks: any other special feature of the accessions is noted.

Accessions are harvested which are considered high yielding by an "eye ball" judgement. Other rows are harvested if they are required for the breeding program or for the maintenance of the germplasm collection.

Step II - Initial Yield Trial (IYT)

The number of entries are now reduced to a more manageable figure. The accessions are further evaluated by one of the following methods:

i.   By growing them in an augmented design (AD)

ii.   By growing them in a replicated row trial (RRT).

If the number of entries is large then an augmented design is followed. No replication is required if the material is grown in an AD, but 3-4 replications are required in an RRT. A row length of 4-5 m is adopted and normal row spacing is followed. For comparison, check entries are included in the trial. The same procedures are followed as described under Step I in respect to cultural practices, nursery note taking, etc. The major departure from Step I is that here the seed yield is actually recorded, and the results are statistically analysed. Superior accessions are then selected for the preliminary yield trials.

Step III - Preliminary Yield Trial (PYT)

Replicated yield trials are conducted, where the number of rows per plot may be 3-4, row length 4-5 m and the number of replications 3-4. The cultural practices, nursery note taking and analysis of results are as in Step II. The superior yielding accessions are selected for evaluation in advanced yield trials.

Step IV - Advanced Yield Trial (AYT)

This trial is dealt with in the same way as that of the PYT, except that the number of rows per replication are 6-8, the row length is 4-6 m, and the number of replications is 4-6. The superior yielding lines are evaluated for quality characters, and only those lines with acceptable seed quality should be further evaluated.

Step V - Adaptation Trial (AT)

Adaptation trials are conducted at different locations representing different agro-environments. These trials are repeated for 2 to 3 years and, at the end, the best 1 or 2 lines are selected.

Step VI - On-farm Trial

The on-farm trial is also known as a 'field verification trial' or a 'demonstration trial'. These trials are conducted on farmers' fields. They are unreplicated, large-plot tests, conducted with 1 or 2 improved lines and the local landrace or cultivars. These trials are conducted at a very large number of sites. The whole range of environments is covered by selecting suitable sites for the conduct of the trials. The objectives of the trial are to grow the new lines under farmers' conditions to assess the yield and to obtain the farmers' views. In most countries an accession is tested in these trials only in one year, while in others accessions are tested over two or three years. If the line is found to be superior in yield and acceptable to the farmers, it is considered for release.

Step VII - Cultivar release

The cultivar is considered for release for commercial cultivation in an area, region, or country, if it is superior to the existing cultivar in yield, or if it has a good yield and is superior for quality or any other special characteristic (such as height, which may facilitate mechanical harvesting). At the time of release of a cultivar it is named, and the results are publicised in popular journals. The seed is increased and distributed to the farmers.

No precise information is available about the exact number of cultivars developed through the direct exploitation of introduced germplasm. However, the number is probably large, and practically all the cultivars developed in the ICARDA region, such as ILC-482 in Syria, Giza 1 in Egypt, Pb 7 in Pakistan and Amdoun in Tunisia, have been developed by this method. Also, the cultivars UC 5 and Mission, in the United States, and Culiacan Cito 860 and Angostura, in Mexico, have all been developed through direct selection. It is clearly a very powerful method of utilizing germplasm.

Utilization of germplasm for the development of cultivars for winter-sowing in the Mediterranean region

Chickpea is grown as a spring-sown crop in the Mediterranean region of west Asia, north Africa, and south Europe, whereas faba bean and lentil are sown in the winter in the same region. Discussions with farmers regarding the sowing of chickpeas during the spring did not reveal the reasons for the adoption of this practice. Therefore, research on spring versus winter sowing was started during the 1974-75 season, which has led to the release, in 1981-1982, of the cultivar ILC 482 for winter sowing in the area of Syria that receives less than 400 mm of rainfall. A brief account of the breeding of ILC 482 is given below:

1974-75: One hundred and ninety two lines of chickpea were sown at Kfardan in the Beqa'a valley of Lebanon. Sub-zero temperatures and snowfall were recorded on several days. The lowest temperature was -12^{0} C. All lines survived the winter, indicating that this crop can tolerate the cold weather of the region.

1976-77: An advanced yield trial, comprising 25 lines, was sown during winter and spring at the University of Aleppo Farm, Muslimieh, Syria. All but two of the winter sown lines were killed by ascochyta blight in contrast to the much higher survival rate of the spring sown lines. This showed why chickpea is not normally sown during winter.

1977-78: A single advanced yield trial and several preliminary yield trials were sown during both winter and spring in north Syria. The crop was protected from ascochyta blight by spraying with fungicide. Yield increases of up to 100% were recorded by winter sowing over spring, indicating the potential of winter sowing.

1978-79: Over 400 genotypes, sown in both winter and spring, were grown in a number of trials with fungicidal protection. The season turned out to be dry, with only 246 mm of precipitation recorded at Tel Hadya. Fungicide spray provided full protection in all but one trial, where even 17 sprays could not save the trial from ascochyta blight. Seed yields of over 3 t/ha were recorded from several lines, confirming the yield potential of the winter-sown crop. It also suggested that it is possible to grow chickpeas in drier areas than is normally the case. The final lesson of the season was that spraying with fungicide was both risky and uneconomic; it is only practical to control ascochyta blight, with resistant cultivars. Therefore, 3000 kabuli germplasm lines were screened for resistance to ascochyta blight, and several sources of resistance were identified.

1979-80: A Chickpea International Yield Trial-Winter (CIYT-W) was initiated with 10 resistant lines, and the trial was conducted at 15 locations in 9 countries. The mean yield of the trial at many locations was 3 t/ha, and the highest yield was up to 4 t/ha. On-farm trials were initiated in Syria; and cultivar, ILC 482, sown during winter, yielded over 100% more than the control, a Syrian landrace which was sown during the spring. Another significant feature was that the yield trials were conducted only with ascochyta blight resistant lines, without any fungicide control.

1980-81: Twenty entries were included in the CIYT-W and sets of the trial were grown in 44 locations in 15 countries. The results were similar to those of 1979-80. The on-farm trial was continued in Syria, and this activity was extended to Lebanon and Jordan.

1981-82: The CIYT-W was further expanded, and 75 sets were distributed to different countries. The on-farm trials were continued in Syria, Jordan and Lebanon and extended to Morocco. 1981-82 was a particularly cold year in northern Syria and many lines suffered badly from cold, and some were killed, not only are ascochyta blight resistant lines needed but also cold tolerant lines. The cultivar, ILC 482, has been released by the Syrian Ministry of Agriculture and Agrarian Reform for sowing during winter in Syria. Another cultivar ILC 484 has been identified for pre-release multiplication in Jordan.

Some of the major conclusions on winter sowing are as follows: First, the seed yield can be doubled by advancing sowing date from spring to winter. Second, winter sowing is possible with lines resistant to ascochyta blight and cold. Third, with the introduction of winter sowing, chickpeas can be grown in drier regions than was previously possible. Fourth, screening of germplasm has resulted in lines resistant to ascochyta blight and cold. Finally, it has been possible to exploit the germplasm to identify lines such as ILC 482 and ILC 484 which have become, or are about to become, commercial cultivars.

Indirect exploitation of germplasm for cultivar development through hybridisation

Utilization of germplasm for specific characters

Several high yielding lines from the chickpea germplasm have been identified, but they lack one or two specific characters, such as disease resistance, insect resistance, cold tolerance or large seed size. With hybridisation, it is not too difficult to add to a high yielding line genes for one or two special characters. For example, a program was initiated four years ago to develop high yielding and ascochyta blight resistant lines. It has been possible to develop several such lines using facilities for off-season generation advancement. In a similar manner, other characters could be added to high yielding lines.

Combining characters to develop better germplasm

The screening of nearly 14000 germplasm accessions, of both desi and kabuli types, for ascochyta blight resistance identified twelve non-location specific sources of resistance. Most of these sources have an intermediate type of seed, are small in size, late in maturity, and are highly photoperiod sensitive. With hybridisation, it has been possible to transfer genes for ascochyta blight resistance to kabuli type germplasm which is less sensitive to photoperiod. A few lines have been identified which have large, kabuli-type seed. These derived lines are better sources of germplasm for breeding programs. Likewise, genes for other characters can be combined, and the resultant genetic stocks could be of more use to plant breeders.

Germplasm maintenance and distribution

It is essential to maintain germplasm accessions at two places to avoid the risk of loss. In the case of chickpeas, duplicate sets are maintained at ICRISAT and ICARDA.

The distribution of germplasm is one of the most important functions of the crop improvement program at ICARDA. Thousands of lines are being distributed through international nurseries at the request of national programs. Every germplasm accession assembled at the center, and every breeding line developed by the center, is available to national programs. The national programs are encouraged to exploit this wealth of material for the benefit of their programs.

12. TAXONOMY, DISTRIBUTION AND EVOLUTION OF THE FABA BEAN AND ITS WILD RELATIVES

J.I. Cubero

Department of Genetics, Escuela Tecnica Superior de Ingenieros Agronomos, Cordoba, Spain

The genus *Vicia* and its main taxonomic features

Vicia can be divided into three different types according to the morphology of the floral peduncle (Figure 10): *ervum*, with a small number of tiny flowers in gracile peduncles (e.g. *V. tetrasperma*); *cracca*, with a high number of flowers in long racemes (e.g. *V. cracca*, *V. villosa*); and *euvicia*, with sessile flowers (e.g. *V. sativa*, *V. narbonensis*), or flowers in small numbers in very short racemes (*V. faba*, *V. melanops*).

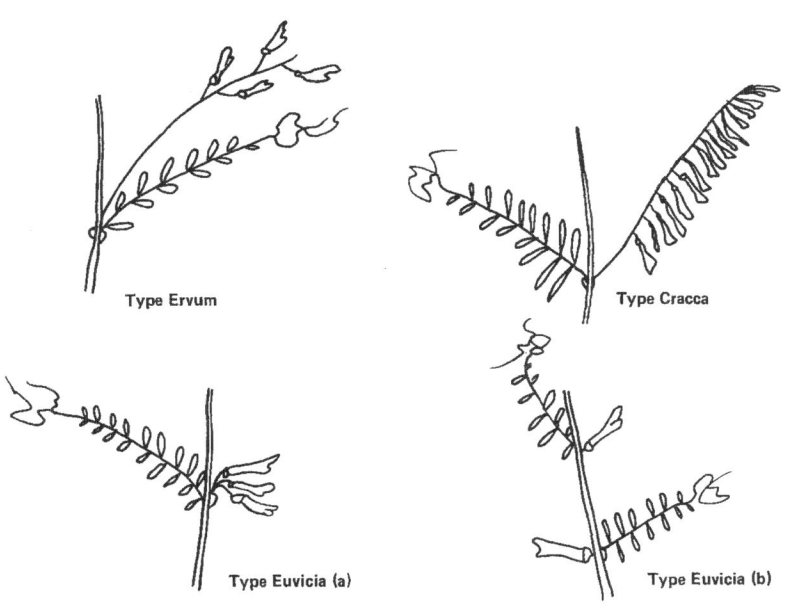

Figure 10. Types of peduncle in *Vicia*.

Of course, these types do not exhaust the variability in floral morphology that can be found in *Vicia*. Thus, in *ervum* and *cracca* the relative size of the peduncle and the rachis is interesting from a taxonomic point of view. Moreover, there is variation in flower size, ranging from very small (4-5 mm in

J.R. Witcombe and W. Erskine (eds.) Genetic Resources and Their Exploitation - Chickpeas, Faba beans and Lentils. ISBN-13:978-94-009-6133-3 (PB)
©1984, Martinus Nijhoff/Dr W. Junk Publishers for ICARDA and IBPGR.

pubescens) to large (more than 20 mm in faba and grandiflora). Ovules per ovary range from 2-4 (pubescens and tetrasperma) to nearly 20 (grandiflora). Other morphological features are also very variable. For example, the floral raceme can be branched (V. aurantiae). The leaflets can range from 2 to more than 25 per leaf and can be minute or extremely large (V. unijuga and V. pseudorobus have leaves even larger than those of V. faba). Seeds are generally rounded, but in some species they are irregular (V. ervilia) or flattened (V. faba, V. michauxii, and some types of V. sativa, such as platysperma). Pods can be cylindrical, or flattened (V. faba, V. sativa) or moniliform (V. ervilia). Nevertheless, the peduncle morphology represented in Figure 10 is useful to assign, at first glance, a specimen to any one of the main taxa of Vicia in any taxonomic system.

There are several such taxonomic systems. The two extreme 'models' are those of the Flora of the USSR and the Flora Europaea, the main features of which are summarised here. The Flora of the USSR recognises three subgenera: Ervilia, Faba and Vicia; the first two are mono-specific, because of the unique morphological features of the species they contain: V. ervilia, with its moniliform pods and irregular-shaped seeds, and V. faba, whose peculiarities are well known to the reader. The third subgenus, Vicia, comprises all the other species of the genus (between 100 and 150), and it is split in sections, cracca, ervum, euvicia, etc., by means of the floral peduncle morphology, size and shape of leaflets, size of flowers and seeds, etc.

At the other extreme, Flora Europaea split Vicia into four sections: cracca, ervum, euvicia and faba. The first three sections are defined according to the morphology of the floral peduncle (Figure 10). Faba includes not only V. faba but V. narbonensis and, on the basis of leaflet size, also V. bithynica. Several other Floras adopt more or less intermediate positions.

The basic chromosome number seems to be n=7, from which n=6 and n=5 have originated. Karyotypes have been described for the same species, differing mostly in chromosomal rearrangements rather than in chromosome number. Polyploidy is, on the contrary, rather rare: less than 10% of the species that have been studied show a chromosome number higher than 2n=20.

Vicia species generally are annual herbs, but there are some perennial ones. Both allogamy (and even functional cleistogamy) are present in the genus, but the numbers of allogamous and of autogamous species are unknown. Partial allogamy is also known (V. faba).

The taxonomic position of V. faba

The taxonomic position of V. faba has been a matter of considerable debate. The two extreme positions are either that there are two distinct genera, Vicia and Faba, or that V. faba belongs to the genus Vicia.   The former viewpoint comes from the Linnean period (the genus Faba was a Tournefort name), and from time to time it is supported by new research data.   In contrast, others argue that V. faba exhibits pure Vicia characters, and there is no reason to place it in a new, distinct genus.   There is an intermediate point of view that I strongly support.   Faba bean is a Vicia species that, since it has several unique characteristics, should be placed in a mono-specific subgenus.   These three positions can be summarised:

i.   Generic rank. Placing faba beans in a separate genus, Faba, different from, but related to, Vicia.   Faba beans would then be named as Faba bona Medikus (Faba vulgaris Moench is a synonym).   The Flora of Iran has recently supported this point of view.

ii.   Sub-generic rank. Placing faba beans as the only species of the subgenus faba of genus Vicia.

iii. Specific rank in section faba of the genus Vicia.   Including faba beans as a species in the genus Vicia, subgenus (or section) faba.   V. narbonensis (and its related V. galilea, johannis, etc.) and, sometimes, even bithynica, would also belong to this subgenus.(The inclusion of bithynica is not easily justified on the basis of classical taxonomy).

Most of the recent taxonomies (e.g. Ball, 1968) adopt the third position, but, in my opinion, this is only because of the lack of systematic studies of this genus, involving morphological, karyological, taxonometrical and chemotaxonomical work.   Subgenus or section faba, as defined in (iii), is commonly based on leaflet size, but one then wonders why amoena, unijuga, dumetorum and many others are also not included.   There are many other arguments against this viewpoint of a homogeneous section faba.

i.   Bueno (1974) studied a collection of living Vicia species by numerical analysis methods; all the possible Euclidean distances between specimens (species or subspecies) were calculated.   Faba beans appeared in this study as an isolated point: the greatest distances found were those between Vicia faba and the other species.   Curiously enough, the other isolated species was V. ervilia; thus, this taxonometric study supported the old Fedchenko (1948) system, the only one splitting Vicia in three subgenera, Faba being one of them.   Fedchenko shows that, from purely morphological considerations, it is also possible to arrive at a unique position for Vicia faba.   Bueno's work proved that the similarity of faba and narbonensis, for example, is more apparent than real.   The only common feature is in leaflet morphology, but such differences as those in floral morphology (in which narbonensis is clearly closer to sativa), and in seed and pod morphology (clearly different from faba), make these two species very distinct.

ii. Chooi (1971 a, and b) studied the variation in DNA nuclear content in the genus Vicia. The study included 45 species belonging to the four sections described by Ball (1968). Giving an arbitrary value of 100 to the DNA nuclear content of Vicia faba, the closest values were those of V. melanops (86) and peregrina (71) (both belonging to the section Vicia), followed by sylvatica (65, a cracca species) and michauxii (62, again a Vicia species). V. narbonensis (55) and bithynica (34) were not close to faba.

The average DNA content per chromosome produced a similar result: V. faba (16.7) and melanops (17.2) had the greatest values, followed by hajastana (11.2) and peregrina (10.2). V. narbonensis (7.8) and bithynica (4.9) were again very far from Vicia faba (Chooi, 1971 a).

Looking deeper into the structure of the DNA of some species, Chooi estimated that about one third of the total DNA was repetitive in faba and narbonensis, but only 15% was repetitive in melanops. Hybrid DNA reassociation experiments indicated that there is little nucleotide divergence among the six species that were studied (Chooi, 1971 b). His studies confirm the special place of V. faba in the genus.

iii. Ladizinsky (1975 a, and b) studied the electrophoretic pattern of V. faba and the narbonensis group (narbonensis, galilea, johannis). The protein profile of the former was completely different from those of the other three species, whose patterns were very similar.

iv. Several authors, reviewed by Cubero (1982), have unsuccessfully attempted to cross V. faba with other Vicia species, particularly with those of the narbonensis group. Crosses with V. faba failed absolutely, whereas those crosses performed between other members of the narbonensis group succeeded.

To cut a long history short, morphological, taxonometrical, electrophoretic, DNA studies and interspecific crosses show that V. faba is not a common Vicia type. If a separate genus is too much (the range of variation includes it in Vicia, perhaps excepting the seed size, clearly an organ men selected for), one more species is not enough. A separate subgenus for it makes sense.

The nearest relatives of V. faba

What are its nearest relatives? According to authors who suggest a subgenus or section faba in the genus Vicia, its nearest relatives are the species included within this subgenus or section. These other species are those belonging to the narbonensis complex (narbonensis, johannis, galilea, etc.) and, sometimes, even bithynica. As has been seen above, crosses between narbonensis and faba have been tried many times, always with the same result: failure. And the taxonometric, as well as the chemical studies just discussed, reveal that, even though they may be morphologically alike, they are really very distant to each other. Their similarity is only in phenotype. In contrast, hybridisation is possible between narbonensis, galilea, etc., indicating that the narbonensis complex is a 'true' group, both on a morphological and a genetical basis.

Table 14. Postulated relatives of V. faba.

---

1. Narbonensis group (on the basis of some morphological traits)
    V. narbonensis L.
        var. narbonensis
        var. aegyptiaca Korn.
        var. affinis Korn.
        var. jordanica Schaf.
        var. salmonea (Mont.) Schaf.
    V. serratifolia Jacq.
    V. johannis Tamanisch.
        var. johannis
        var. procumbens Schaf.
        var. ecirrhosa (M. Pop.) Schaf.
    V. galilaea Plit. & Zoh.
        var. galilaea
        var. faboidea (Plit. & Zoh.) Schaf.
        (probably V. hyaniscyamus Mout. = V. galilaea)
2. V. melanops L. (on the basis of DNA content)
    V. peregrina L. (id).
3. V. bithynica L. (on the basis of some taxonomic keys)

Those thinking that V. faba deserves at least a section of its own, obviously consider it a species without close relatives. Future studies will tell if this is true or not. Sooner or later, the wild ancestor of V. faba should be found and we will then know if it has close relatives or if it is an isolated species. It would not be the only case, even among food legumes. Meanwhile, a different way to choose species to cross with faba beans is possible. This is to look for species with a similar DNA content to that of V. faba. As has been seen above, narbonensis has about half of this content, bithynica about a third, and most of the Vicia species do not reach more than 60%. Only two surpass this level: melanops, with almost 90%, and peregrina, with 75%. Both species are morphologically very different to V. faba. Crosses may be impossible for many other reasons: similarity in DNA content is not the only factor allowing successful hybridisation, but it will be interesting to try them. Of course, we know the DNA contents of only a fraction of the genus and there may be other species also meeting this karyological requirement. To solve the problem of the nearest relatives of V. faba it is essential to have information on much larger collections of Vicia species than those now existing. It is a very important point to test not only a large number of species but many different populations within species; crossability frequently has a genetic basis.

The postulated V. faba relatives are summarised in (Table 14 above), and they include V. bithynica because of its presence in the subgenus (or section) faba of some taxonomic treatments of Vicia.

Distribution of possible wild relatives of V. faba

V. narbonensis is a circum-Mediterranean species and, in fact, its habitat extends far beyond these limits. The rest of the species of the narbonensis group are restricted to the Near East, which indicates that this is the possible place of origin of V. narbonensis. Even though the group has merited some monographic studies, it is not well known genetically. Collecting expeditions in the Near East are really very important; only knowledge accumulated over the last few years has allowed the description of new species, and consequently narbonensis has been divided into several species.

V. melanops is a typical European Vicia, extending from the Mediterranean regions to central Europe. V. peregrina is a very common weed in European wheat fields.

Figures 11 to 14 show the distribution maps for these species.

Taxonomy of V. faba

We will not consider here the classifications of the last century.

Muratova (1931), a disciple of Vavilov, performed the first serious systematic studies on the taxonomy and evolutionary aspects of this species early this century. She worked with excellent collections, now unfortunately lost, and probably she was the last person observing authentic primitive populations on a large scale. In fact, her system is still often used. She defined two subspecies, based on the maximum number of leaflets per leaf: faba (eufaba in her nomenclature), with more than four leaflets per leaf, and paucijuga, with less than four as a maximum. Subspecies faba was split into three botanical varieties according to the coefficient of thickness/length of the seed: major (flattened seeds), equina (medium) and minor (rounded seeds). Every variety was divided into groups (grex), forms, etc.

Hanelt (1972) reviewed this system in the last decade. He also recognised

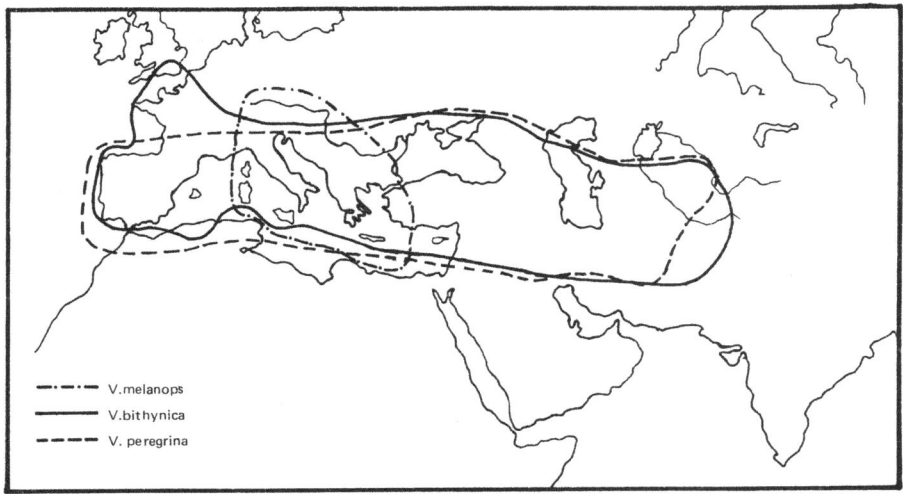

Figure 11. Distribution of V̲. melanops, V̲. bithynica and V̲. peregrina.

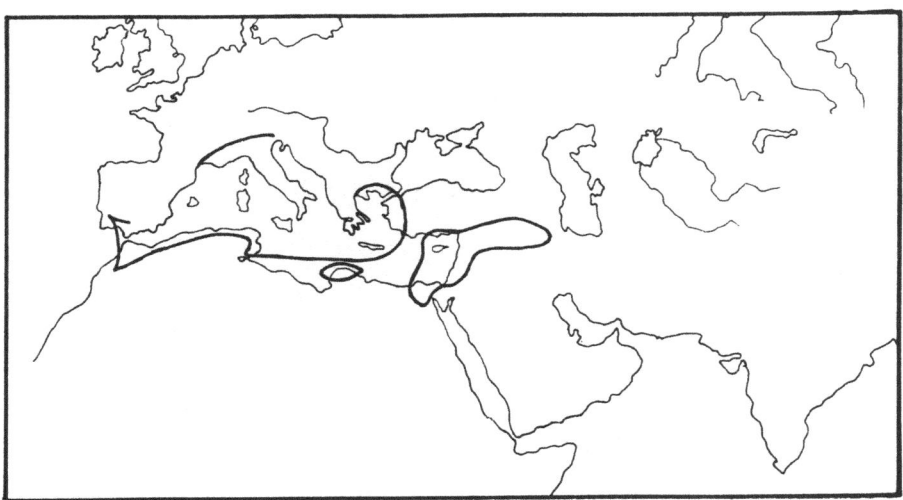

Figure 12. Distribution of V̲. narbonensis. (After Schafer, 1973).

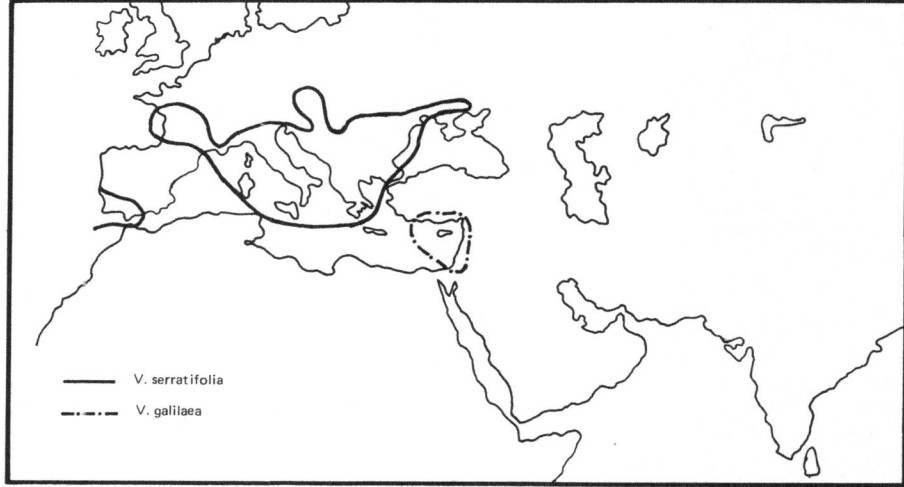

Figure 13. Distribution of <u>V</u>. <u>serratifolia</u> and <u>V</u>. <u>galilaea</u> (After Schafer, 1973).

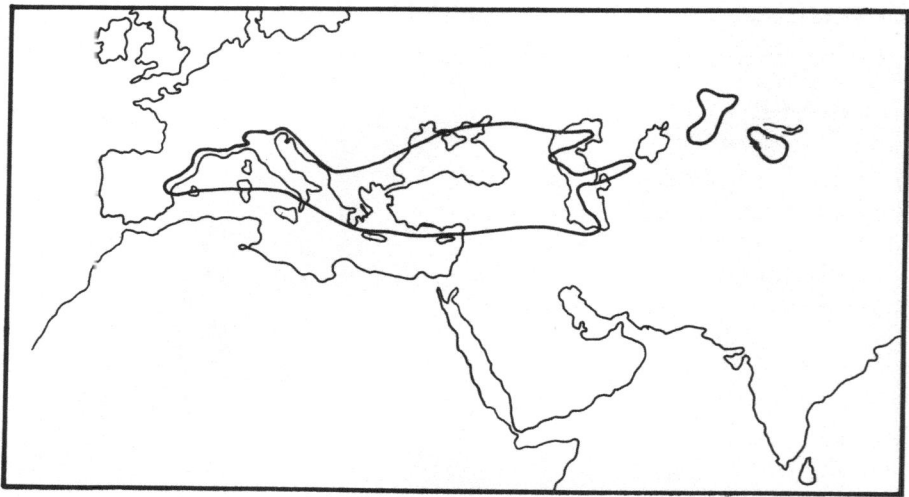

Figure 14 Distribution of <u>V</u>. <u>johannis</u>. (After Schafer, 1973).

two subspecies, faba and minor, including in the latter the paucijuga of Muratova. The principle for the division is the same: the two subspecies are recognised on the basis of their morphological differences.

For almost fifteen years I have been working with plants and populations belonging to the four Muratovian main names (paucijuga, minor, equina and minor). For both theoretical and practical purposes we have crossed them in all directions, finding that crossability is similar between groups and that it depends more on specific genotypes than on 'subspecies' or 'varieties'. In fact, our only problems arose within accessions of minor, probably because of the presence of some incompatibility genes. Now, if we accept the biological concept of species, based on the possession of a common gene pool, we have to admit that the existence of two subspecies is rather a morphological artifact. Following the system proposed by Harlan and de Wet for crop species taxonomy, all the known cultivated forms of V. faba must be classified in only one subspecies: V. faba faba (subspecies A) (Figure 4). Subspecies B, which would include the wild forms, is empty. The four Muratovian names (major, equina, minor and paucijuga) are considered by us as only simple groups or grex. They are useful because they identify, with just a single word, a set of characteristics; but in relation to the biological concept of species, they are almost meaningless.

These facts do not imply that there are not differences between primitive paucijuga forms (not all the accessions classified as paucijuga are really primitive) and the rest of the faba forms. The study of segregating populations from crosses between primitive paucijuga and modern forms have shown that paucijuga contains recessive genes for most of the vegetative characters and dominant genes for most of the pod and seed characteristics, excepting the seed shape: the rounded (ellipsoidal) seed shape of paucijuga (and of minor) is recessive, and the flattened seed shape of equina and major is dominant (all these characteristics are polygenic).

Evolution

Evolution in Vicia

This is a question linked to that of the evolution of Vicieae. All the Vicieae genera form a 'continuum' with border species difficult to assign and name correctly.

The evolution in Vicieae has only merited fragmentary studies. In the cases of Pisum and Lens there are more or less viable hypotheses, but these are not thoroughly proven. Lathyrus remains unstudied from this viewpoint. In Vicia, studies on a very small number of examples (some in the 'sativa' complex) show that structural chromosomic rearrangements can play a role in the speciation of

Vicia.

The origin of V. faba is unknown. How can we explain its DNA content of twice the average of the other Vicia species? Has V. faba the primitive DNA content of Vicia ancestors?   It is currently thought that this 'double' in information is arranged in a linear way in the chromosome, not in a multi-stranded way.   But it has to be demonstrated that the DNA content of V. faba is really the duplication of a primitive genotype.   In fact, even though it is true that V. faba has twice the DNA average of the other Vicia species, there is a continuous distribution of the DNA content in the genus.

Intraspecific evolution

Faba bean is a crop of the late Neolithic period. It is a crop of the 'first industrial revolution' - the Bronze Age. It appeared suddenly during the fifth millenium B.C. in the Near East, in the 'Fertile Crescent', the cradle of agriculture.   Archaeological remains have been discovered in many Mediterranean and European regions, but many important zones have yet to be explored (for example, Iran and Afghanistan).   However, data concerning faba beans fit very well with the pattern of many other crops such as chickpea.   Faba beans spread from the Fertile Crescent to different regions, following the steps of the 'first generation' crops (barley, einkorn, emmer, lentils).   However, the spreading of the 'second generation' domesticates (faba bean, chickpea) was more rapid: trade routes were increasing in number because of the 'first metal fever';the search for tin to obtain bronze. Thus, it reached the eastern coast of Spain by the end of the millenium, and later several points in Central Europe, mainly by way of the Danube. Far eastern countries (China and Japan) extensively cultivated faba beans only after 1,200 A.D., introductions taking place, very probably, by the Silk Road.   Most of the forms reaching the far eastern regions were of the major type, that had been selected by the end of the Roman period (5th - 6th centuries A.D.).

Because most of the primitive forms known at the present time were collected in Afghanistan and northern India, this region has also been proposed as the centre of origin of the faba bean. This would lead to the acceptance of the existence of people domesticating plants in those remote regions. A lot of exploration has to be done there, but this suggestion goes against the accepted theory of the origins of agriculture. The important fact is, that even in our century, paucijuga-like forms were found in countries far away from Afghanistan. Muratova described as Vicia pliniana a very primitive type found in Algeria. The difference from V. faba was mostly anatomical (I have never been able to see any difference between pliniana and primitive forms of V. faba in herbarium specimens).

There are two possible explanations for the presence of paucijuga forms of faba beans in northern India and Afghanistan, which do not necessitate the hypothesis that this area was the crop's 'centre of origin'. The first explanation involves introgression between wild and cultivated forms in the east, and the second is dependent on the selection and fixation of genes in the domesticated crop:

i.   Small genetic differences between local eastern and western populations of the wild ancestor of V. faba would have produced differences in the morphology of the local cultivars. This would be possible if the area of distribution of the wild ancestor extended from Turkey to the Afghan mountains, and there was introgression between the crop and its wild relative.

ii.   Genetic drift and selection, which would be favoured by the topography of the Afghanistan region, would fix for ever the genetic characteristics of the primitive, early domesticated material. This primitive material could have been transported eastwards early in the history of domestication, and may have been of a narrow genetic base and resembled paucijuga.

In the absence of new discoveries, the first explanation is the most plausible. In any case, in the east (northern India and northern Pakistan, and Afghanistan) some cultivated material of really primitive phenotype was maintained. Theses forms are so primitive in phenotype, dehiscent and lacking in self-fertility, that we had thought that we were facing the wild ancestor. But the forms with the most primitive (wild) characteristics that we have studied in our collection are indehiscent and very self-fertile, indicating obvious human selection.

As with many other crops, faba beans reached Ethiopia soon after domestication. Most of the samples collected there are small seeded (botanically minor); this fact could indicate that when the new crop colonised Ethiopia, new faba bean types had not yet been produced. But the Ethiopian forms, being minor, are not primitive paucijuga forms like those found in the eastern wing of the faba bean geographical distribution.

After the Ethiopian 'conquest', a new and potentially important seed form was obtained: the flattened seed. Our studies show that this shape is controlled by a polygenic system. For this reason, I do not think that it was produced by a single mutation, but rather by the accumulation of minor genes by unconscious selection. To me, the evolution of varieties with flattened seeds, that is, the equina type, is a crucial point in the evolution of faba beans. To increase the seed size by the growth of the ellipsoid minor seeds in a regular way (i.e., all the three dimensions growing at the same rate) would result in problems for the pod structure, particularly if the pod has mechanical tissues for dehiscence. In contrast, growth in only two dimensions (length and width) enables an almost continuous growth in the size of the seed, and hence of both the seed weight and the total yield (the high correlation between seed size and yield is well known: major types are the most productive, even considering their unstable yield pattern).

The genetics of yield and seed characters support this hypothesis: the low paucijuga yield is recessive, but both its high number of pods and seeds are dominant. The flattened shape of the seed is dominant over the rounded one. Obviously, the primitive farmers did not select for a high number of pods or seeds. or for flattened or rounded seeds, but for yield. A significant increase in yield, however, was not possible before the selection of the first mutants with flattened seeds. From this moment on, minor genes for flattened seeds (that is, for most cellular divisions in the length-width plane) were being accumulated as an indirect response to the selection pressure for yield. All the same, a correlated response to increased seed size produced a decrease in both the number of pods and seeds per plant. The raw material for the production of the modern types of faba bean was thus obtained.

Summarising: the ancient farmers, selecting for higher yield, obtained types with flattened seeds (equina) by unconscious accumulation of genes affecting the pattern of cell division of the seed. The new seed shape, that is the equina forms, permitted higher yields and, therefore, were favoured by the farmers. This is why I have considered equina as the 'central nucleus of evolution', not because it was the primitive form but because it maintained genetic potential to produce new forms.

The last form to evolve was the major type. This was probably obtained as a consequence of the continuous selection pressure to increase yield or, its strongly correlated character seed size, which is equivalent to extremely flattened seeds. Archaeologically, the first remains of authentic major seeds go back to the 10th century. We can easily imagine that it first appeared some 500 years before, that is, during the late Roman period. We do not know anything about the date of the discovery of indehiscent pods, or whether the discovery of indehiscency occurred only once or several times with several independent genes controlling this character. The actual distribution of this character is circum-Mediterranean, as is to be expected, since indehiscency is not strongly selected for in humid regions. Anyway, it is not illogical to suppose that the evolution of the major types was in many cases associated with the selection of indehiscent, fleshy pods, able to contain very big seeds. Perhaps the very refined Roman cuisine pressed gardeners to select these edible forms out of equina.

References

Ball, P.W. 1968. Flora Europaea 2 Pages 128-136. eds. T.G. Tutin et al., London: Cambridge University Press.

Bueno, M.A. 1974. Taxonometria y cariologia en el genero Vicia. Ph. D. Thesis, Madrid, Spain.

Chooi, W.Y. 1971 a. Variation in the nuclear content in the genus Vicia. Genetics 68: 195-211.

Chooi, W.Y. 1971 b. Comparison of the DNA of six Vicia species by the method of DNA-DNA hybridization. Genetics 68: 213-230.

Cubero, J.I. 1982. Interspecific hybridization in Vicia. Pages 91-108 in Faba Bean Improvement, eds. G. Hawtin and C. Webb, The Hague: Martinus Nijhoff.

Fedchenko, B.A. 1948. Vicia. In Flora of the URSS, ed. V.L. Komorov, Moscow.

Hanelt, P. 1972. Die infraspezitische Variabilitaet von Vicia faba und ihre Gliederung. Kurturpflanze 20: 75-128.

Harlan, J.R. and de Wet, J.M.J. 1971. Towards a rational classification of cultivated plants. Taxon 20: 509-517.

Ladizinsky, G. 1975 a. On the origin of the broad bean, Vicia faba L. Israel Journal Botany 24: 80-88.

Ladizinsky, G. 1975 b. Seed protein electrophoresis of the wild and cultivated species of section faba of Vicia. Euphytica 24: 785-788.

Muratova, V. 1931. Common beans (Vicia faba). Bulletin of Applied Botany, Genetics and Plant Breeding, Supplement 50.

Further reading

Cubero, J.I. 1973. Evolutionary trends in Vicia faba. Theoretical and Applied Genetics. 43: 59-65.

Cuberc, J.I. 1974. On the evolution of _Vicia faba_. Theoretical and Applied Genetics 45: 47-51.

Cuberc, J.I. and Suso, M.J. 1981. Primitive and modern forms of _Vicia faba_. Kulturpflanze 29: 137-145.

Zohary, D. 1977. Comments on the origin of cultivated broad bean, _Vicia faba_ L. Israel Journal of Botany 26: 39-40.

Zohary, D. and Hopf, M. 1973. Domestication of pulses in the Old World. Science 132: 887-894.

13. GENETIC RESOURCES OF FABA BEANS

J.R. Witcombe

FAO/IBPGR, South West Asia Program

Introduction

The derivation of Vicia faba is unclear, and its immediate ancestor is not known.   Since V. faba does not produce fertile hybrids with any other species, its gene pool is restricted to itself.   The use of V. narbonensis, or other Vicia species, as sources of genes for V. faba improvement using conventional breeding techniques is considered unlikely (Cubero, Chapter 12).

With regard to variability, the most comprehensive work is that of Muratova (1931). However, her study was restricted to a few basic traits. No one region can be designated as a centre of diversity, and material from Europe, N. Africa, S.W. Asia, India and China are equally important.   Germplasm materials from outside of this area, e.g. north and south America and Australia, are likely to be of recent origin and therefore not primitive cultivars.   Such advanced cultivars are usually part of breeders' working collections and, as such, are not threatened by genetic erosion.

Existing collections

Table 15. Holders of major collections of V. faba

Institute	Country	No. Accessions
ICARDA, Aleppo	Syria	1931 V. faba
N.I. Vavilov IAPI, Leningrad	USSR	2525 V. spp.*
GL, CNR, Bari	Italy	1469 V. faba
ZGK, Gatersleben	GDR	786 V. faba
SVP, Wageningen	Netherlands	700 V. faba
PBRICL, Tumenice	Czechoslovakia	500 V. faba
IPP, Braunschweig	FRG	804 V. faba

* All Vicia spp. Vicia faba numbers unknown.

The IBPGR is developing a network of Plant Genetic Resources Centres, each with the responsibility for maintaining the world collection of a particular crop. No institute has yet accepted this responsibility for faba beans. However, ICARDA has one of the largest faba bean collections in existence, even

J.R. Witcombe and W. Erskine (eds.) Genetic Resources and Their Exploitation - Chickpeas, Faba beans and Lentils. ISBN-13:978-94-009-6133-3 (PB)
©1984, Martinus Nijhoff/Dr W. Junk Publishers for ICARDA and IBPGR.

though physical facilities to house this material under ideal conditions are not yet completed.

In 1980, the IBPGR published a Directory of Germplasm Collections of Food Legumes (Ayad and Anishetty, 1980). Holders of major collections of faba bean germplasm listed in this publication are summarised in Table 15. This list is, doubtless, incomplete and will need future revision.

The germplasm collections held at ICARDA, Braunschweig and Bari are summarised in Table 16. It is obvious from these data that certain countries although having indigenous varieties of V. faba, are seriously under-represented. Examples are Iran, China and India.

The numbers of accessions from different countries are correlated between the collections. There must be a great deal of duplication between them even though, at least in the case of the Braunschweig and ICARDA collections, there has been no direct seed exchange between them.

Data describing existing collections are incomplete. In most germplasm collections, many accessions are found in which even the country of origin is not recorded. Because of this problem, and the incompleteness of evaluation data, it is not possible to describe the main characteristics of the germplasm material according to region of origin. Although such information would be useful and interesting, it is largely unavailable in the case of all crop plants.

Collecting expeditions

Recent collecting expeditions have taken place in Cyprus, Egypt and Afghanistan. Over the past few years, collections of V. faba have been made in the Mediterranean region by the Germplasm Laboratory, Bari, Italy, and in Ethiopia by the Ethiopian Genetic Resources Centre.

The collections made by Bari include Algeria in 1976 (47 accessions), Greece, Spain and Tunisia in 1977 (with 23, 31 and 18 accessions, respectively), and Crete in 1978 (38 accessions).

In Cyprus in 1980, the Agricultural Research Institute, Cyprus, in collaboration with FAO and IBPGR, collected 98 accessions of V. faba (Della, 1981). Ninety-five collections were made in Egypt in 1978 by the International Institute of Tropical Agriculture (Badra, 1978), and M.M.F. Abdalla (pers. comm.) has made extensive faba bean collections in Egypt; the major one being in 1979, when 200 samples were collected from all regions of Egypt. Duplicates have been sent to Braunschweig for long term storage. In Afghanistan in 1974, fifty collections of V. faba were made (Solh et al., 1974).

It is certain that this is not a complete list of all recent collecting expeditions for the faba bean. I know of some collecting trips that have collected very few faba bean samples.   Other, more extensive, collections of faba bean have been made, but the material has not been distributed outside of the country of collection.   There must also be collecting expeditions of which I am unaware.

ICARDA is attempting to stimulate, with assistance from the IBPGR, the collecting of faba bean germplasm in India and various other regions where collections are under-represented.

Multiplication and rejuvenation of $\underline{V}$. $\underline{faba}$ germplasm

The partially cross-pollinated habit of $\underline{V}$. $\underline{faba}$ has profound genetic consequences, and affects the method of maintaining its germplasm.   Before examining two simple genetic models of the consequences of out-crossing, we must first define some terms.

Out-crossing is the rate at which individual plants cross-pollinate, or open-pollinate, to other plants (instead of selfing). $\underline{V}$. $\underline{faba}$ shows a high degree of out-crossing.   Figures for open-pollination in faba beans vary widely but maxima of 70% (Holden and Bond, 1960) and 61% (Poulsen, 1975) have been reported.   However, in most circumstances, out-crossing is likely to be in the range of 5 to 50%.

Inter-crossing is the rate of pollination between different accessions (entries) of germplasm material.   The result of inter-crossing is that, after a number of generations, the identity of the individual accession is lost.

Hence, the rate at which the identity of accessions is lost, is directly related to the rate of inter-crossing.   In turn, this depends on both the rate of out-crossing and the degree of isolation between entries.   An extreme example of lack of isolation would be when each plant of an accession was totally surrounded by foreign material, so that the rate of inter-crossing would equal that of out-crossing.   In practice, since the plants of an accession are always planted together in rows or plots, inter-crossing will always be less than out-crossing.   It is worth noting that the worst planting pattern from the viewpoint of inter-crossing is single, long, one-plant-width rows of each accession, with a small distance between rows.

Normally the maximum possible rate of inter-crossing between any two accessions will be about half the rate of out-crossing, since at least 50% of a plant's neighbours will be members of its own accession.   We then have a maximum rate of inter-crossing of 50% (i.e. half of the maximum out-crossing rate of 100%).

We can consider a simple model in which two accessions differ completely for a single locus so that the initial difference in gene frequency will be 1.   Even in the extreme case of 50% inter-crossing, it will take six generations for the

Table 16. Origin of accessions at ICARDA, Bari, Braunschweig (BGRC) and Gatersleben (ZGK) and area under production in various countries.

| Country | No. of accessions | | | | Area* under |
	ICARDA	Bari	BGRC	ZGK	faba beans 1981(1000 ha)
Afghanistan	98	72	13	6	n.d.+
Algeria	21	34	-	-	46
Argentina	1	-	1	-	1
Australia	2	-	4	2	n.d.
Austria	1	-	-	1	n.d.
Bangladesh	2	-	-	-	n.d.
Belgium	-	-	-	1	n.d.
Bolivia	1	-	1	-	11
Brazil	-	-	-	-	173
Bulgaria	-	1	-	4	n.d.
Canada	2	122	-	-	n.d.
China	9	7	-	6	2200
Colombia	14	-	-	-	n.d.
Cyprus	-	103	-	-	3
Czechoslovakia	-	-	8	25	41
Dominican Republic	-	1	-	-	9
Ecuador	13	-	-	-	8
Egypt	57	85	211	28	105
Ethiopia	370	95	211	73	325
Finland	11	-	1	4	n.d.
France	9	10	2	15	21
Germany DR	-	-	-	11**	6
Germany FED	257	-	17	12**	4
Greece	25	55	6	43	6
Guatemala	-	-	-	-	20
Holland	33	19	3	15	n.d.
Hungary	10	66	-	2	n.d.
India	9	6	1	3	n.d.
Iran	13	9	4	2	n.d.
Iraq	57	52	5	1	16
Italy	48	247	12	151	162
Japan	5	4	4	-	1
Jordan	18	3	1	-	n.d.

Country	No. of accessions				Area* under faba beans 1981(1000 ha)
	ICARDA	Bari	BGRC	ZGK	
Lebanon	30	30	-	-	(1)
Libya	-	-	-	-	7
Mexico	1	-	1	-	46
Mongolia	-	-	-	1	n.d.
Morocco	15	31	109	12	130
Nepal	1	-	2	-	n.d.
Pakistan	7	3	1	-	n.d.
Paraguay	-	-	-	-	16
Peru	2	-	1	4	23
Poland	12	1	4	71	n.d.
Portugal	5	5	-	1	31
Romania	-	-	16	-	n.d.
Spain	77	107	6	103	63
South Africa	1	-	-	-	n.d.
Sri Lanka	2	-	-	-	n.d.
Sudan	35	22	-	4	17
Sweden	10	4	1	1	n.d.
Switzerland	1	-	3	3	n.d.
Syria	62	32	4	3	8
Tunisia	49	54	2	1	77
Turkey	120	72	19	16	30
UK	88	56	11	32	(40)
Uruguay	1	-	-	-	n.d.
USA	2	2	1	-	n.d.
USSR	21	6	21	47	n.d.
Yemen	6	26	18	-	n.d.
Yugoslavia	14	17	10	6	n.d.
N. Europe	82	-	-	-	n.d.
Unknown	199	2	69	30***	n.d.
TOTAL	1929	1461	804	786**	

*   From FAO Production Year Book, 1981.

**   + 48 with origin as 'Germany'.

***   Includes 7 from Latin America and 8 mutants.

+   n.d. = no data available.

two accessions to become identical (Figure 15).   Nevertheless, after three
generations of such inter-crossing, the difference in gene frequency at this
locus is reduced from 1 to 0.1.

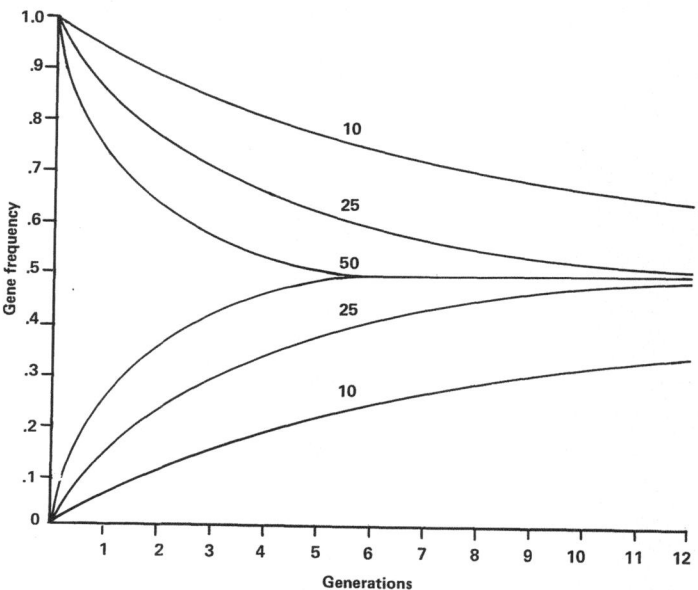

Figure 15. The rate at which two populations, initially differing for a gene, become
identical at 50, 25 and 10% inter-crossing.

In the case of faba beans, gene flow will always be much less than 50%, since
rates of out-crossing are lower than 100%, and planting patterns should be
designed so that there will always be greater separation between the accessions
than where 50% of a plant's neighbours are foreign. Progress towards genetic
identity at 10% and 25% inter-crossing is shown in Figure 15.   It can be seen
that when inter-crossing is reduced, it takes many generations for the
accessions to become identical.

Another situation is an accession which differs from all others by uniquely
having allele A at a particular locus. We assume that this accession '10' has
the unique allele 'A' at a frequency of 1 and inter-crosses with the other
accessions at a rate of 10%. After one generation of inter-crossing, the allele
'A' would be found in other accessions and could be present in the foreign
pollen that inter-crosses with accession '10'. This possibility is disregarded
since it is present at such a low frequency.   Under the above conditions, the
allele is lost slowly from the population (Figure 16). It is still present at a
frequency of 0.1 after 21 generations, but is in great danger of being lost due

to random drift. Moreover, if it is recessive, it is becoming increasingly hard
to recover, as after 20 generations double recessives only comprise 1% of the
genotypes in accession '10'.

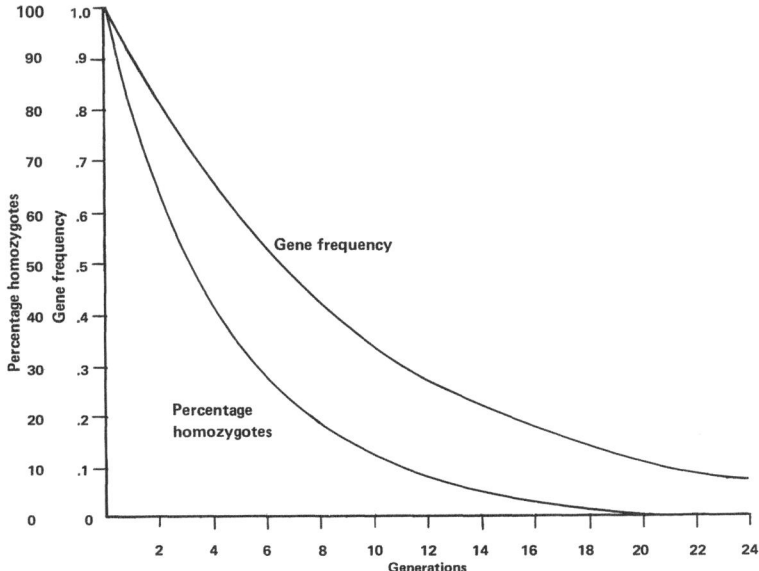

Figure 16. The rate at which a population loses a gene when all other populations are different at that locus, at 10% inter-crossing.

It can be seen from the models that, although the affects of inter-crossing
are profound, in the absence of random drift they occur slowly at the levels of
inter-crossing to be found in V. faba. A certain degree of inter-crossimg can
be tolerated in germplasm collections without losing the identity of the
individual lines.

As a consequence of the partial out-crossing of V. faba, there are three main
ways of maintaining it in germplasm collections; as populations, inbred lines,
and trait-specific gene pools.

Accessions maintained as populations

When the germplasm collection is maintained as populations, steps are taken
to reduce the rate at which the collections become identical. This is done by
reducing inter-crossing and by reducing generation advance over time.

Reducing inter-crossing between accessions

Despite the open pollination shown by faba bean, germplasm material can be successfully maintained as collections of populations.

Growing plots isolated by distance only, and discarding the border plants, is an effective method of reducing inter-crossing. Under Syrian conditions with only 13 m between the plots, inter-crossing is around 10%.

The size of the plots was very important and an increase of 12 m^2 (75%) in plot area had a large effect (Table 17). This is an unfortunate fact when it is considered that often only small seed samples are available, so that plot sizes are small.

Table 17. Inter-crossing in faba beans. (Adapted from Hawtin and Omar, 1980 a).

	Mean inter-crossing between plots (%)		
	border	centre	whole
Plot size 16 m sq.	14.7	10.1	14.2
Plot size 28 m sq.	9.5	6.7	8.9
Average inter-plot distance 11.0 m	12.3	10.4	11.6
Average inter-plot distance 14.3 m	11.8	6.1	10.9

To overcome this problem, small plots of V. faba can be surrounded with a species that does not inter-cross with it but attracts the same pollinating insects. In this season at ICARDA, plots of V. faba are being surrounded by oil-seed rape (Brassica campestris) to see if the pollinating insects first visit these border plants and lose their foreign faba bean pollen to them. A physical barrier of taller plants, e.g. Triticale, is also being tried.

The dispersal of pollen by visiting insects will not bear a simple relationship to distance, so that equal increases in isolation distance will not have an equal effect in rates of inter-crossing. If there is a sigmoid relationship between distance and rate of inter-crossing, which is likely, then there is a critical distance above which the plots should be planted. Any further increase in this distance will not have a large effect.

The data of Porceddu et al. (1980) indicate that if basal pod clusters are discarded, then inter-crossing will be reduced by a factor of a half. These authors found that rates of inter-crossing between varieties was 26% in the first to fifth pod clusters and only 16% in the clusters above. These results are supported by Hanna and Lawes (1967) and Poulsen (1975).

Reducing generation advance

It is clear that with inter-crossing, the more generations that an accession is grown, the more it loses its genetic identity. The effects of inter-crossing are reduced by minimising generation advanced using careful seed management (Figure 17). Early generation seed (foundation seed) is maintained in medium or long term storage and used conservatively by removing sub-samples from it only for the purpose of multiplication. The multiplication seed is then used for the same collection for distribution and evaluation purposes. Only when this multiplied seed is exhausted or lost, is a further sample withdrawn from the foundation seed (Figure 17). This scheme is a more conservative method than the simple one of using the base collection, or active collection, as a source. If the base collection is so used then generation advance may be accelerated, for the base collection will need to be regenerated when its quantity falls to a minimum level.

Ultimately, generation advance cannot be eliminated, and is dependent on the longevity of the seed in the base collection. Fortunately, V. faba seed is very long lived.

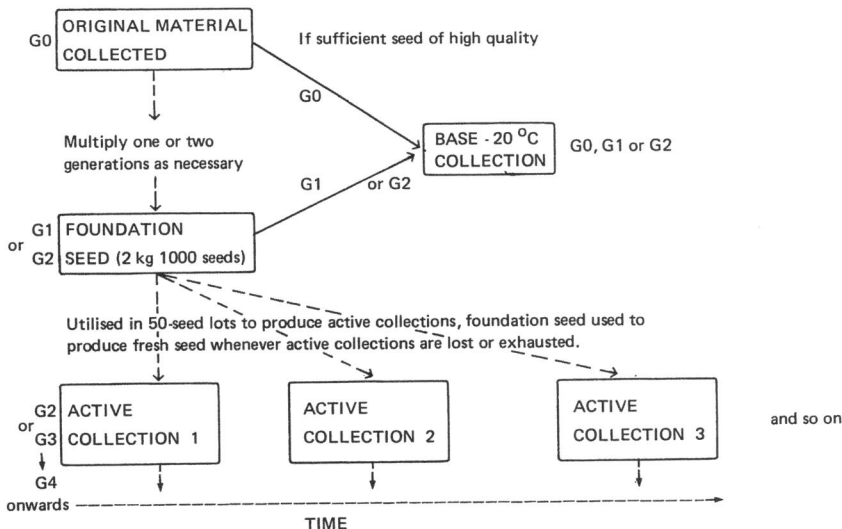

Figure 17. Idealised flow diagram for reducing generation advance in out-crossing germplasm material maintained as open-pollinated populations.

Germplasm collections maintained as inbred lines

Burton (1979) has argued that selfing (by bagging) is a good way of maintaining cross-pollinated germplasm. There are many arguments in its favour.

i.   For V. faba, selfing by bagging (or growing plants in a screen house in the absence of pollinating insects) requires less work than deliberate inter-crossing between plants of an accession, and is less likely to produce out-crossing by error to other accessions.

ii.   After several generations, inbred lines are formed which retain their genetic identity from generation to generation. The inbred lines can be planted in rows to provide replicate plants to facilitate screening.

iii. Selfed seed will retain most of the genes from the original open-pollinated population. Due to genetic drift resulting from the small sample size and random segregation, some genes that occur in the open-pollinated population will not be found in the inbred lines derived from it. This factor becomes less important as the sample size is increased by producing more inbred lines per accession. However, the more variable an accession is, the greater the sample size required to guard against gene loss. In any case, recessive lethals, and those genes which are closely linked to them, will be lost.

iv.   Selfing uncovers genes. Material is then available for immediate screening for recessive gene characteristics. Recessive genes at low frequency in an open-pollinated population will mostly be found in heterozygotes and are therefore hidden.

Hawtin and Omar (1980 b) describe pure line collections of V. faba maintained at ICARDA. From one to four single plants are selected from the original accession, depending on its heterogeneity, and progeny rows from these selected plants are grown in the subsequent season in a screen house. The lines are then maintained, each generation, from a single representative plant taken from the progeny row. Although some genetic variation is lost when single plants are selected from the original accession, it has the advantage, described above, that the identification of desired characters, particularly those controlled by recessives, is more efficient.

When inbred lines are produced, it is still necessary to keep base collections (and possibly foundation seed) of the open-pollinated populations from which they were derived. Further pure lines can then be produced, if desired, from the base collection or early generation seed. The inbred lines themselves are maintained in base collections for long term storage, so that they can more easily be replaced if they are lost from the active collections.

Burton (1979) also describes a method of using selfed seed to simplify the distribution of open-pollinated germplasm. To avoid having to measure and ship several hundred packets of germplasm, he created and registered Tift No. 1 S_1 pearl millet germplasm.   This is a mixture of equal quantities of selfed seed from 275 pearl millets, and contains a great variety of the dominant and recessive genes observed in the original open-pollinated populations from which it was derived.   All characters can be observed in a single space planting of a 25 g sample (c. 5000 seeds).   To distribute seed, only one 25 g sample is required, instead of 275 labelled packets.

One disadvantage of Burton's method is that no pedigree data are available to the breeder who, therefore, has to base his selections on single plants, which is a problem for characters of low heritability.

Trait specific gene-pools

Germplasm pools can be prepared by mixing seed of the accession and growing them together in an isolated field.   The plants are allowed to inter-cross naturally and are harvested as a bulk.   This is not the best way to preserve germplasm.   Burton (1979) found that advancing five germplasm pools of pearl millet three generations narrowed the phenotypic variability of the original pool, lost genes, and obscured 'hard-to-recover characteristics'.

Loss of phenotypic variability can be overcome by dividing the germplasm material into different pools on the basis of morphological characters (traits) and region of origin.   The first step is to evaluate the accession for characters with reasonably high heritability.   The results of evaluation are used to create trait specific gene-pools (TSG's) based on such characteristics as maturity (early, late), seed size, height and growth habit.   The TSG's are formed from accessions with the same country or region of origin since they will tend to be similar.   This also ensures that only accessions from similar latitudes with similar daylength sensitivity are pooled.   A high correlation for flowering time between evaluation sites can then be safely predicted when the different sites in which the TSG's are subsequently grown and evaluated have a similar daylength.

Multivariate analysis of evaluation data would provide the best basis of forming TSG's.   Clusters of accessions in the analysis can then be pooled on the basis of a whole range of characters and the correlations between them. Multivariate analyses successfully separate accessions into geographic groups (Murphy and Witcombe, 1981).

Each TSG thus consists of a number of selected accessions which are mixed in equal proportions and maintained as a bulk by growing them as a single isolated population with natural inter-crossing.

The total phenotypic variance is maintained between the germplasm pools and a collection of around 2000 accessions could be reduced to 10-200 TSG's.   Within each germplasm pool, linkages are broken so that within gene pool variance and total phenotypic variance may even increase. Recessive gene characters are more likely to be exposed within each trait specific gene-pool than when a single gene-pool is created from all of the material.

For germplasm purposes, TSG's are best maintained by subjecting them to minimum selection pressures (e.g. constant seed descent, where the same number of seeds are taken from each plant in the gene-pool). TSG's can then be distributed, and used as source populations for such things as improvement by single plant selection, recurrent selection and mass selection. By allowing natural selection to take place, adaptation to local conditions is improved.

It must be pointed out that trait specific gene-pools can consist of inter-crossing populations, or mixtures of inbred lines, or selfed seed. The former method has a lower frequency of homozygous loci, making screening for recessive characters more difficult, but has the advantages of allowing gene recombination and being easier to maintain.

Gene pooling has distinct advantages when it is realised that variability is distributed non-randomly throughout the distribution of a crop plant. Combining all early maturing varieties from Syria into a single gene-pool is not likely to lose much variability, but it can reduce the numbers of accessions enormously. A major problem with germplasm material is that it can be very repetitive, with little difference between many of the accessions, and trait specific gene-pools exploit this fact to reduce accession numbers drastically.

Gene-pools, as in the case of inbred lines, do not avoid the necessity of maintaining, in base collections, the original populations from which they were derived. The original populations are the source material from which the gene-pools are assembled, and contain the entire range of genes.

Storage conditions and seed size

To increase longevity, seeds are stored at low moisture and low temperature. Under the long term storage conditions recommended by the IBPGR of 5-6% moisture and -20°C, the longevity of V. faba seeds is very high (Roberts, 1975). For example, under these conditions, seed samples will take decades to fall from 100% to 85% germination. 'Medium term' storage of low moisture content, and temperatures of 5°C to 10°C still provides very high longevity periods. Active collections are kept at ambient temperatures, and the moisture content is allowed to equilibrate naturally. Depending on the ambient conditions, seed germination can remain high for up to five years. Problems will only be encountered in warm, humid climates.

The seed size of V. faba (up to nearly 2000 g per 1000 seeds) precludes the storage of samples with a large number of seeds. However, the arbitrary and impracticable figure, for large seeded species, of 12000 seeds per accession for heterogeneous accessions, which is recommended by the IBPGR, can safely be ignored; whatever the circumstances, a considerable portion of the variability will be maintained between the accessions. Furthermore, large sample sizes of individual accessions are only necessary if a useful gene occurs at a low frequency in a single accession, and no example of this can be found in the genetic resources literature. Useful genes are likely to occur in more than one accession, and often at a high frequency.

Variation, evaluation and utilization

Chapman (1981) has drawn up a list of the genetic variation within V. faba which is available to breeders. Many of the forms described are spontaneous or induced mutations, whereas some have been selected from landrace populations of cultivars. This catalogue should prove valuable to the breeder, and provide stimulus for further work.

The preponderance of mutants, rather than forms from landrace populations, in this list, probably reflects the work that has been done. Inbreeding of landrace populations to reveal recessive genes and produce inbred lines has revealed useful genetic variation. The lack of useful genetic variability for pest and disease resistance may show a need for further screening, rather than a lack of variability for these characters.

There is a need to specify conditions under which plants are screened for genetic variation, and to agree on methods of evaluation. A draft list of descriptors is shown in Appendix 10, pending the publication of an IBPGR/ICARDA list.

Evaluation data can be divided into two broad categories, preliminary and full. Preliminary evaluation data, including characterization, is produced by genetic resources centres, and consists of a short list of descriptors to provide some preliminary knowledge of the germplasm collections. Such preliminary evaluation data can be of immediate use, as it is essential in the pooling of germplasm material. Breeders, when screening germplasm material, will produce evaluation data that would normally lie outside the scope of preliminary evaluation. Information on biochemical quality and race-specific disease resistance would enter into this category.

Acknowledgements

I would like to acknowledge M.M.F. Abdalla of Cairo University, and the late Ali Abdel Aziz of the Agricultural Research Centre, Egypt, for sending me details of Egyptian faba bean germplasm. Lothar Seidewitz sent me a listing of the faba bean collection at Braunschweig, Pietro Perrino that of Bari, and Christien Lehman that of Gatersleben. Their help is gratefully acknowledged. Geoff Hawtin gave many helpful suggestions at the manuscript stage.

References

Ayad, G. and Anishetty, M. 1980. Directory of Germplasm Collections 1. Food Legumes. Rome: IBPGR.

Badra, T. 1978. Egypt. Pages 123-125 in Genetic Resources Unit Exploration 1978. Ibadan, Nigeria: IITA.

Burton, G.W. 1979. Handling cross-pollinated germplasm efficiently. Crop Science 19: 685-690.

Chapman, G.P. 1981. Genetic Variation Within Vicia faba. FABIS: ICARDA.

Della, A. 1981. Broad beans collecting in Cyprus. Plant Genetic Resources Newsletter 44: 17-19.

Hanna, A.S. and Lawes, D.A. 1967. Studies on pollination and fertilisation in the field bean (V. faba L.). Annals of Applied Biology 59: 289-295.

Hawtin G.C. and Omar, M. 1980 a. Estimation of out-crossing between isolation plots of faba beans. FABIS 2: 28-29.

Hawtin, G.C. and Omar, M. 1980b. International faba bean germplasm collection at ICARDA. FABIS 2: 20-22.

Holden, F.H.W. and Bond, D.A. 1960. Studies on the breeding system of field beans (V. faba L.). Heredity 15: 175-192.

Muratova, V. 1931. Common beans -Vicia faba L. Supplement 50th Bulletin Applied Botany: 1-298.

Murphy, P.J. and Witcombe, J.R. 1981. Variation in Himalayan barley and the concept of centres of diversity. Pages 26-36 in Barley Genetics IV: Edinburgh University Press.

Porceddu, E., Monti, L.M., Frusciante, L. and Valpe, N. 1980. Analysis of cross-pollination in _Vicia_ _faba_ L. Zeitschrift Pflanzenzuechtung 84: 313-322.

Poulsen, M.H. 1975. Pollination, seed-setting, cross-fertilisation and inbreeding in _V._ _faba_ L. Zeitschrift Pflanzenzuechtung 74: 97-118.

Roberts, E.H. 1975. Problems of long-term storage of seed and pollen for genetic resources conservation. Pages 269-295 _in_ Crop Genetic Resources for Today and Tommorrow. eds. O.H. Frankel and J. G. Hawkes IBP, 2. Cambridge: Cambridge University Press.

Solh, M., Rashid, K. and Hawtin G.C. 1974. Food legume Collection, Afghanistan, July-August, 1974. Mimeograph. Ford Foundation: ALAD.

Appendix 10. A list of faba bean descriptors.

PASSPORT DATA

1. ACCESSION DATA   (As per IBPGR standard format)

2. COLLECTION DATA   (As per IBPGR standard format)

CHARACTERIZATION AND PRELIMINARY EVALUATION DATA

3. SITE DATA   (As per IBPGR standard format)

4. PLANT DATA
    4.1 VEGETATIVE
         4.1.1 Seedling stem pigmentation
         4.1.2 Plant height
         4.1.3 Leaflet size
         4.1.4 Branching
    4.2 INFLORESCENCE AND FRUIT
         4.2.1 Time to flower
         4.2.2 Time to maturity
         4.2.3 Flower ground colour of standard petal (flag)
         4.2.4 Pattern of streaks on standard petal (flag)
         4.2.5 Wing petal colour
         4.2.6 Pod shape
         4.2.7 Pod colour at maturity
         4.2.8 Pod surface reflectance
         4.2.9 Pod angle
         4.2.10 Pod length
         4.2.11 Number of seeds per pod
    4.3 SEED
         4.3.1 Testa ground colour
         4.3.2 Hilum colour
         4.3.3 Seed shape
         4.3.4 100 seed weight

FURTHER CHARACTERIZATION AND EVALUATION

5. SITE DATA   (As per IBPGR standard format)

6. PLANT DATA
    6.1 VEGETATIVE
        6.1.1 Leaflet shape
        6.1.2 Number of leaflets per leaf
        6.1.3 Stipule spot pigmentation
        6.1.4 Stem thickness
    6.2 INFLORESCENCE AND FRUIT
        6.2.1 Pod indehiscence
        6.2.2 Number of flowers per inflorescence
        6.2.3 Height of lowest pod
        6.2.4 Autofertility
    6.3 SEED
        6.3.1 Testa pattern
        6.3.2 Seed yield
        6.3.3 Vicine and convicine content
        6.3.4 Cooking time

7. STRESS SUSCEPTIBILITY
    7.1 Low temperature
    7.2 High temperature
    7.3 Drought
    7.4 High soil moisture
    7.5 Winter kill
    7.6 Salinity

8. PEST, DISEASE AND WEED SUSCEPTIBILITY
    8.1 PESTS
        8.1.1 Aphids (_Aphis_ spp.)
        8.1.2 Leaf weevils (_Sitona_ spp.)
        8.1.3 Stem borer (_Lixus_ spp.)
        8.1.4 Seed weevils (_Bruchus_ spp.)
        8.1.5 Leaf miners (_Liriomyza_ spp.)
        8.1.6 Stem nematode (_Ditylenchus_ _dipsaci_)
        8.1.7 Others
    8.2 FUNGI
        8.2.1 Chocolate spot (_Botrytis_ _fabae_)
        8.2.2 Ascochyta blight (_Ascochyta_ _fabae_)
        8.2.3 Leaf spot (_Alternaria_ spp.)
        8.2.4 Rust (_Uromyces_ _fabae_)
        8.2.5 Powdery mildew (_Erysiphe_ _polygoni_)
        8.2.6 Root rot complex (_Rhizoctonia_ spp., _Fusarium_ spp., etc.)

14. STRATEGIES FOR EXPLOITING THE FABA BEAN GENE POOL

G.C.Hawtin
ICARDA, Aleppo, Syria

Introduction

Pollinating system

Numerous reports from Europe have indicated that from 5-80% cross-pollination occurs in faba beans under field conditions. The majority of estimates fall in the range 20-50% (e.g. Fyfe and Bailey, 1951; Sjodin, 1977). Although comparatively few estimates of cross-pollination have been reported in the Middle East region, the situation appears similar. Kambal (1969) reported 35% cross-pollination near Khartoum, Sudan, and El-Sherbeeny (1970) found between 57 and 71% at Giza, Egypt.

In Europe, the long-tongued bumble bees (e.g. Bombus hortorum and B. agrorum) appear to be the main pollen vectors (Sjodin, 1977). Honey bees are generally considered less effective (Bond and Hawkins, 1967), although they can adequately pollinate the crop if present in sufficient numbers (Poulsen, 1973). In the Nile Valley, honey bees may be more important; in one study in Egypt, about 80% of the insects visiting faba bean flowers were honey bees (Wafa and Ibrahim, 1957).

Autofertility

It has been known for many years that insect pollination in faba beans can result in increased seed set. On entering the flower an insect may cause the release of the stamens and style from the keel petal (a process known as tripping) and thereby help the transfer of self pollen from the anthers to the stigma.

Autofertile lines do not have a tripping requirement (Drayner 1956, 1959) and lines with high levels of autofertility have been widely reported (e.g. Hanna and Lawes 1967; Poulsen 1979). Recently, several breeding programs have concentrated on developing autofertile cultivars (Lawes, 1980). Such cultivars should have the advantage of being able to yield well even in the absence of adequate insect pollinators, resulting in improved yield stability. Research on autofertility has led to the release of several cultivars, Dacre, Danas, and Deiniol, by the Welsh Plant Breeding Station (Bond, 1979).

Although autofertility has been reported in lines of European origin it may be more prevalent in paucijuga populations from India and materials originating in Africa or the Mediterranean area (Lawes, 1980). The Sudanese line Ik (Kambal, 1969), a 'Mediterranean type' (Hayes and Hanna, 1968) and the Egyptian

J.R. Witcombe and W. Erskine (eds.) Genetic Resources and Their Exploitation - Chickpeas, Faba beans and Lentils. ISBN-13:978-94-009-6133-3 (PB)
©1984, Martinus Nijhoff/Dr W. Junk Publishers for ICARDA and IBPGR.

cv. Habashy (Hanna and Lawes, 1967) have all been identified as highly autofertile. From a study of 188 germplasm accessions originating from 19 countries, it was concluded that the most autofertile lines were those which originated in Ethiopia, Egypt, Iraq and Syria (Filippetti, 1979). However, not all of the material originating from west Asia and north Africa is autofertile. The Egyptian cvs. Giza 1, Giza 2 and Rebaya 40, and the Sudanese Baladi have all responded to tripping (Kambal 1969, Hanna and Lawes, 1967).

Drayner (1959) reported that inbreeding results in a loss in autofertility in inbred lines but that this is fully restored in the F_1 between them. This finding has been confirmed subsequently by several researchers, although selection for autofertility within and between inbred lines has been successful (Poulsen, 1977; Toynbee-Clark, 1971; Lawes 1973). The greater autofertility of F_1's may, along with other factors, be associated with an increased production of pollen (Drayner, 1956).

Breeding objectives

Increased green or dry seed yield and improved yield stability are the primary objectives of most faba bean breeding programs. The low heritability of the characters, and consequent limited genetic advance for yield in response to selection, has led many scientists to search for characters which are associated with yield but which are more highly heritable. As in the case with most grain legumes, correlations between the components of yield and yield itself are frequently large. However, compensation between components can severely limit yield gains in response to selection for single components. A greater chance of success in indirect selection for yield might come from selecting for various phenological or morphological attributes. Characters such as the onset of flowering, duration of flowering, onset of grain filling, duration of grain filling, number of tillers and, possibly, plant height might all be used in the construction of selection indices for the improvement of yield (de Vries, 1979).

Breeding for increased yield, however, should perhaps be bolder. If faba beans are to compete economically with other crops, substantial yield advances are necessary, and soon. The exploitation of heterosis through synthetic and, ultimately, hybrid cultivars, or a substantial remodelling of the plant's growth habit, to transform the current indeterminate habit to a determinate one, are two such approaches which could pay off in improved yield potential; while increasing autofertility, reducing flower drop, and developing resistance or tolerance to major yield-reducing factors (pests, diseases, environmental stress, etc.) could all improve the stability of yield.

Insufficient attention has been paid in the past to developing cultivars with improved disease or pest resistance. Botrytis fabae, for example, has long been recognised as a major cause of low yields in certain environments, yet very few sources of resistance to the disease have been identified.   In recent large-scale screening of inbred lines, however, the situation looks more promising (Hanounik and Hawtin, 1982) and the lack of success in the past may well be due to inadequate screening methods being applied to too small a range of genotypes.   Delaying the planting date has frequently been associated with reduced yields in Mediterranean-type environments but early planted crops may be more heavily attacked by certain pests and diseases, such as Orobanche, ascochyta blight or chocolate spot.   In Sudan, early planting, to escape the effects of high temperatures at the end of the season, can lead to an increased incidence of seedling root rot/wilt diseases (Freigoun, 1980).   The development of resistance might enable the crop to be sown earlier in such environments, which, in turn, could lead to substantial yield increases.

Other breeding objectives, such as improved cooking or nutritional quality, increased nitrogen fixation, tolerance to environmental stresses and improved characteristics for mechanisation, are all considered to be important by certain programs.

Genetic variation

Considerable genetic variation has been reported within the species (Abdalla, 1976; Bianco et al., 1979; Porceddu et al., 1979) and germplasm resources are still largely unexploited.   The development of appropriate screening techniques for the identification of desired genotypes is of great importance; these tests should ideally be quick, reliable and non-destructive, to enable a large number of populations, lines or single plants to be screened.   In faba beans, seedling screening techniques are particularly valuable, in that selected plants can subsequently be isolated from insect pollinators to ensure selfing.

At ICARDA, a set of pure lines, selected from within each accession of the base collection, is maintained for initial screening purposes (Hawtin and Omar, 1980; Witcombe, Chapter 13).   The pure lines are considered of particular value in that:

i.   Screening can be carried out on several plants of the same genotype, e.g. on a row basis.

ii.   If destructive techniques are used the specific genotypes selected are still available.

iii. Recessive genes, which may be 'buried' in heterozygous accessions, can be identified.

The maintenance of a pure-line collection, however, requires resources beyond those of most national programs: over one hectare of land is under insect-proof screen mesh at the Tel Hadya site near Aleppo, Syria.   Small samples of entries in the ICARDA pure-line collection, however, can be made available to faba bean breeders on request.

The full exploitation of the available genetic resources may require hybridisation to allow gene recombination. This may either be carried out by means of insect pollinators or, for greater control, by hand-crossing.

Artificially induced mutation has also proved to be a valuable technique for creating genetic variation in faba beans (e.g. Sjodin, 1971; Nagl, 1979). Its use in the future may also give rise to useful genes, especially when desired traits cannot be identified within existing germplasm collections. Polyploidy and inter-specific hybridisation may also play a role in future breeding programs. Their potential fully justifies the considerable research effort which may be required before practical techniques for their exploitation can be developed (Cubero, 1982).

Types of commercial cultivars

Cultivars which breeders are currently developing fall into four main categories:

i.   Open-pollinated populations.
ii.   Synthetics.
iii.   Hybrids.
iv.   Fully autogamous lines.

Although most of the commonly grown cultivars are open-pollinated populations, synthetics are becoming increasingly important, especially in northern Europe. Hybrids are not yet produced commercially, and further research is necessary before they can be developed as an economic alternative. Both synthetics (Bond, 1982) and hybrids (Picard et al., 1982) exploit heterosis, which may result in both greater and more stable yields than those of non-autofertile open-pollinated populations.

The question arises as to how far inbreeding should be utilized in a breeding program. The situation with respect to synthetics has been considered by Bond (1982), and it would appear that well-evaluated inbred lines are the most promising components of such cultivars. When the end product is an open-pollinated cultivar the value of inbreeding is less clear.

It has been proposed that faba beans could, or should, be developed as a fully autogamous species (e.g. Lawes, 1980). A conversion of the breeding system to autogamy not only offers an opportunity for improved yield stability (through less dependence on pollen vectors), but would allow populations and segregants to be screened and multiplied under open pollination. This in turn would greatly increase both the scale on which breeding operations could be conducted and the degree of selection pressure which could be applied (Bond,

1979). The closed flowered character (Poulsen, 1977), coupled with cleistogamy, could prove important in the development of fully autogamous cultivars. Although autogamy would exclude the exploitation of heterosis, except possibly through the use of male sterility, the numerous advantages of such a system fully justify the expenditure of considerable research effort.

Evaluation and selection methods

Mass selection, either within local populations or in generations following hybridisation, is probably the most widely used breeding method, but it has given rise to little improvement in yield when selection has been made on the basis of yield itself (Sjodin, 1977).   It is of greatest value for improving characters of high heritability, especially if these characters are strongly correlated with yield.   However, the main value of mass selection may lie in improving uniformity (Bond, 1971).

Recurrent selection has proved a useful technique for the improvement of many allogamous species, and various schemes have been proposed for its use in partially allogamous grain legumes such as faba beans (Rachie and Gardner, 1975).   Such methods may have particular value in the improvement of characters which are under polygenic control, as is the case with yield and many of the characters of importance in faba bean breeding.   A system of recurrent selection, using honey-bees in cages to randomly inter-cross flowers in the recombination phase, is currently being investigated at ICARDA.   A similar system is being used in Egypt (Nassib et al., 1979).   Recurrent selection can give rise either to improved populations for direct release as open-pollinated cultivars, or can provide a source of genetic variability for further improvement by other methods, including the development of inbred lines.

At ICARDA, the main breeding strategy is to develop, through isolation in segregating generations, relatively homogeneous populations with desired agro-economic characteristics. Selfing is ensured through the use of individual plant pollinating bags in early segregating generations, and screen house and isolation plots in more advanced generations.

In this scheme, materials are selected in early generations (by mass selection/single plant selection) on the basis of highly heritable characters; in particular disease and pest resistance. In later generations, selection for yield assumes greater importance when there is adequate seed for replicated, and ultimately, multi-location trials.

Single seed descent, as proposed by Brim (1966), carried out in an insect-free environment can offer an efficient means of advancing populations without selection, while preserving genetic variability for later selection. The method is of particular interest if it is possible to raise two or more generations per year under these conditions.

The use of a pedigree, or bulk-pedigree breeding system without adequate facilities for selfing is particularly dangerous in faba beans, especially for the improvement of yield. The selection of superior individuals is likely to result in the selection of many F_1 out-crossed plants, which exhibit heterosis.

A major obstacle to the improvement of faba beans, especially major types, is the low multiplication rate, frequently less than 10-15 fold. Research is needed on field plot designs, particularly micro-plot techniques, that can be used by breeders to evaluate lines comprising small numbers of seeds.

Prospects

It is expected that new breeding techniques will be developed in the future which could have a profound effect on the genetic improvement of faba beans. These might include techniques which are still in their infancy, such as interspecific, or intergeneic hybridisation, protoplast culture, pollen culture, or even genetic engineering. However, it is doubtful whether these approaches will ever completely replace traditional breeding methods, and, in any case, faba bean breeders cannot await such developments.

Unfortunately, there is still no general agreement between breeders as to which breeding methods offer the most promise, and indeed there probably is no ideal solution. In spite of this, if the crop is to increase or even maintain its role in world agriculture, it is important that its productivity and reliability be substantially improved.

References

Abdalla, M.M.F. 1976. Natural variability and selection in some local and exotic populations of field beans, Vicia faba L. Zeitschrift Pflanzenzuechtung 76: 334-343.

Bianco, V.V., Damato G., Miccolis, V., Polignano, G., Porceddu, E. and Scippa, G. 1979. Variation in a collection of Vicia faba L. and correlation among agronomically important characters. Pages 217-250 in Some Current Research on Vicia faba in Western Europe, eds. D.A. Bond, G.T. Scarascia-Mugnozza and M.H. Poulsen, Luxembourg: EEC.

Bond, D.A. 1971. Breeding methods in field beans (Vicia faba L.). Pages 119-126 in Report of the Meeting of EUCARPIA Fodder Crops Section, Lusignan, 1970.

Bond, D.A. 1979. Breeding work on _Vicia_ _faba_ in the U.K. FABIS 1: 5-6.

Bond, D.A. 1982. Development and performance of synthetic varieties of _Vicia_ _faba_ L. Pages 41-51 _in_ Faba Bean Improvement, eds. G. Hawtin and C. Webb, The Hague: Martinus Nijhoff.

Bond, D.A. and Hawkins, R.P. 1967. Behaviour of bees visiting male sterile field beans (_Vicia_ _faba_). Journal of Agricultural Science, Camb. 68: 243-247.

Brim, C.A. 1966. A modified pedigree method of selection in soybeans. Crop Science 6: 220.

Cubero, J.I. 1982. Interspecific hybridisation in _Vicia_. Pages 99-108 _in_ Faba Bean Improvement, eds. G.Hawtin and C. Webb, The Hague: Martinus Nijhoff.

Drayner, J.M. 1956. Regulation of out breeding in field beans. Nature 177: 489-490.

Drayner, J.M. 1959. Self- and cross-fertility in field beans (_Vicia_ _faba_ Linn.). Journal of Agricultural Science, Camb. 53: 387-403.

El-Sherbeeny, M.H. 1970. Studies on pollination, fertilization and pod-setting in the field bean and their bearing on breeding the crop. MSc. Thesis. Univ. of Cairo.

Filippetti, A. 1979. Breeding projects and work for the improvement of broad beans (_Vicia_ _faba_) in Puglia. Pages 168-188 _in_ Some Current Research on _Vicia_ _faba_ in Western Europe, eds. D.A. Bond, G.T. Scarascia-Mugnozza and M.H. Poulsen, Luxembourg: EEC.

Freigoun, S.O. 1980. Effect of sowing date and watering interval on the incidence of wilt and root rot diseases in faba beans. FABIS 2: 41.

Fyfe, J.L. and Bailey, N.J.T. 1951. Plant breeding studies in leguminous forage crops. I. Natural cross-breeding in winter beans. Journal of Agricultural Science, Camb. 41: 371-378.

Hanna, A.S. and Lawes, D.A. 1967. Studies on pollination and fertilization in the field bean (_Vicia_ _faba_ L.). Annals of Applied Biology 59: 289-295.

Hanounik, S. and Hawtin, G.C. 1982. Breeding for resistance to chocolate spot caused by Botrytis fabae. Pages 243-250 in Faba Bean Improvement, eds. G. Hawtin and C. Webb, The Hague: Martinus Nijhoff.

Hawtin, G.C. and Omar, M. 1980. Estimation of out-crossing between isolation plots of faba beans. FABIS 2: 28-29.

Hayes, I.D. and Hanna, A.S. 1968. Genetic studies in field beans. III. Variation in self-fertility in a diallel cross. Zeitschrift Pflanzenzuechtung 60: 315-326.

Kambal, A.E. 1969. Flower drop and fruit set in field beans, Vicia faba L. Journal of Agricultural Science, Camb. 72: 131-138.

Lawes, D.A. 1973. The development of self-fertile beans. Pages 163-176 in Welsh Plant Breeding Station Report 1972.

Lawes, D.A. 1980. Recent development in understanding, improvement and use of Vicia faba. Pages 625-636 in Advances in Legume Science, eds. R.J. Summerfield and A.H. Bunting: H.M.S.O.

Nagl, K. 1979. Results of mutation and breeding work on Vicia faba in Austria. Pages 355-369 in Some Current Research on Vicia faba in Western Europe, eds. D.A. Bond, G.T. Scarascia-Mugnozza and M.H. Poulsen, Luxembourg: EEC.

Nassib, A.M., Ibrahim, A.A. and Khalil, S.A. 1979. Methods of population improvement in broad bean breeding in Egypt. Pages 176-178 in Food Legume Improvement and Development, eds. G.C. Hawtin and G.J. Chancellor, Ottawa: ICARDA-IDRC.

Picard, J., Berthelem, P., Duc, G. and Le Guen, J. 1982. Male sterility in Vicia faba. Future prospects for hybrid varieties. Pages 53-69 in Faba Bean Improvement, eds. G. Hawtin and C. Webb, The Hague: Martinus Nijhoff.

Porceddu, E., Bianco, V.V., Damato, G., Miccolis, V. and Polignano, G. 1979. Variability of some agronomical characters in 158 Italian accessions of Vicia faba L. Pages 251-265 in Some Current Research on Vicia faba in Western Europe, eds. D.A. Bond, G.T. Scarascia-Mugnozza and M.H. Poulsen, Luxembourg: EEC.

Poulsen, M.H. 1973. The frequency and foraging behaviour of honey bees and bumble bees on field beans in Denmark. Journal of Apiacal Research 12: 75-80.

Poulsen, M.H. 1977. Obligate autogamy in Vicia faba L. Journal of Agricultural Science, Camb. 88: 253-256.

Poulsen, M.H. 1979. Performance of inbred populations and lines of Vicia faba L. ssp. minor. Pages 342-354 in Some Current Research on Vicia faba in Western Europe, eds. D.A. Bond, G.T. Scarascia-Mugnozza and M.H. Poulsen, Luxembourg: EEC.

Rachie, K.O. and Gardner, C.O. 1975. Increasing efficiency in breeding partially outcrossing grain legumes. Pages 285-297 in International Workshop on Grain Legumes: ICRISAT.

Sjodin, J. 1971. Induced morphological variation in Vicia faba L. Hereditas 67: 155-180.

Sjodin, J. 1977. Methods of breeding broadbeans (Vicia faba). Pages 148-161 in Food Legume Crops: Improvement and Production. Rome, Italy: FAO Plant Production and Protection Division.

Toynbee-Clarke, G. 1971. Pollination studies with highly-inbred lines of winter beans (Vicia faba L.). Journal of Agricultural Science, Camb. 77: 213-217.

de Vries, A.P. 1979. In search of characters to be used for indirect selection on grain and protein yield in Vicia faba L. Pages 324-341 in Some Current Research on Vicia faba in Western Europe, eds. D.A. Bond, G.T. Scarascia-Mugnozza and M.H. Poulsen, Luxembourg: EEC.

Wafa, A.K. and Ibrahim, S.H. 1957. Temperature as a factor affecting pollen-gathering activity by the honey bee in Egypt. Bulletin, Faculty of Agriculture, Ain Shams University, No. 162.

15. EVALUATION AND UTILIZATION OF FABA BEAN GERMPLASM IN AN INTERNATIONAL BREEDING PROGRAM

F.A. Elsayed

Food Legume Improvement Program, ICARDA, Aleppo, Syria

Introduction

Faba beans (Vicia faba L.) have been one of the main sources of protein for people in the Middle East and north Africa since ancient times. Today, faba bean is becoming even more important as a source of protein; faba beans are not only high in protein content (up to 35%) but are also an excellent source of lysine and supplement a cereal-based diet which is deficient in this amino acid.

In 1980, faba beans were grown on 0.8 million hectares in the Middle East and north Africa (FAO, 1980) and produced a total of 796 thousand tonnes, which corresponds to an average seed yield of less than one tonne/ha. On the other hand, in the same year in Europe average seed yields ranged from 1.5 to 3.0 tonne/ha. There would appear to be a great potential for increasing faba bean production in ICARDA's region.

This chapter outlines the evaluation and utilization of faba bean germplasm at ICARDA from the viewpoint of plant breeding.

The germplasm collection

The germplasm collection of faba beans at ICARDA consists of primitive forms, landrace populations, genetic stocks, breeding lines and released cultivars. The major use of the germplasm collection is to provide a source of genes for traits such as high yield, yield stability, wide adaptation, disease and insect resistance, improved plant type, and improved physiology (reduced flower drop), etc. For further details on the germplasm collection see Witcombe (Chapter 13).

J.R. Witcombe and W. Erskine (eds.) Genetic Resources and Their Exploitation - Chickpeas, Faba beans and Lentils. ISBN-13:978-94-009-6133-3 (PB)
©1984, Martinus Nijhoff/Dr W. Junk Publishers for ICARDA and IBPGR.

Utilization

The method of utilization of faba bean germplasm in ICARDA's breeding program is shown in Figure 18. Pure lines (BPL) are developed from each accession of the collection (ILB), which has the great advantage that recessive genetic variation is revealed and characters are fixed. The disadvantages in producing pure lines are their lower yield and stability and the loss of alleles from the original accessions due to genetic drift. Moreover, the production of the BPL collection requires large amounts of capital and labour, since the self-pollination has to be done under an insect-proof screen.

Approximately 1000 BPL lines were evaluated for various morphological and agronomic characters at Tel Hadya in 1980-81 (Table 18). Table 19 lists the 16 traits that are being scored in the evaluation of the BPL collection.

Table 18. Range of some agronomic and morphological traits found in 1000 faba bean pure line (BPL) accessions evaluated at Tel Hadya in the 1980-81 season.

Trait	Range*	BPL No. for preferred extreme
Total no. of flowers/plant	16-220	398, 847, 1294, 1339.
Total no. of green pods/plant	2-42	44, 172, 179, 762.
Total no. of mature pods/plant	2-22	977, 1283, 1623, 1731.
Seed yield (g/m^2)**	6-650	148, 225, 471, 924.
Plant height (cm)	30-115	1466, 1538 (short);144, 966 (tall).
Stem width (mm)	2.0-9.3	1256, 1267, 1339, 1781.
No. of branches/plant	1-7	1313, 1654, 1699, 2059.
No. of leaflets/leaf	2-8	470, 877, 1093, 1380.
Leaflet width (cm)	1.5-6.8	290, 304, 794, 1711.
Time to 50% flowering (d)	75-125	1429, 1041 (early);161, 733 (late).
Protein content (%)	19.8-34.4	331, 521, 563, 717

*   Mean of three randomly selected plants/plot.
**   Middle row data.

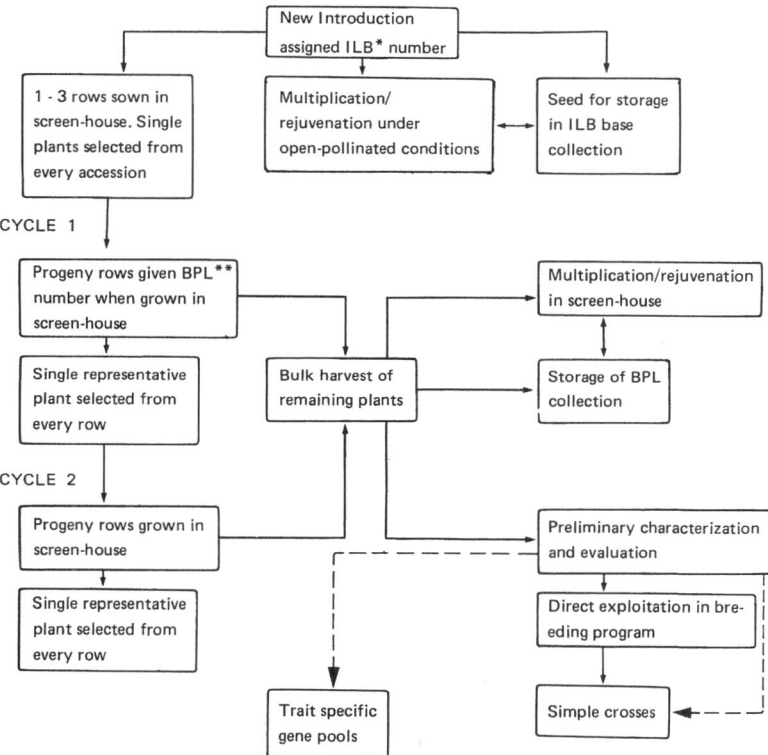

* International Legume Bean
** Bean Pure Line

Figure 18. A flow chart for the maintenance and evaluation of faba bean germplasm at ICARDA.

Table 19. Characters evaluated in the BPL collection.

1.	Time to 50% flowering (d)
2.	Time to end of flowering (d)
3.	Flower colour
4.	Total number of flowers per plant
5.	Leaflet width (mm)
6.	Number of leaves on nodes 12-15
7.	Number of leaflets per leaf on nodes 12-15
8.	Number of green pods per plant
9.	Plant height (cm)
10.	Number of branches per plant
11.	Number of reproductive nodes per plant
12.	Height above ground of lowest pod (cm)
13.	Stem width (mm)
14.	Lodging index
15.	Pod dehiscence index
16.	Pod angle

Breeding for yield

Introduction

One aim of the ICARDA faba bean breeding program is to identify, evaluate and develop sources of high-yielding, broadly-adapted genotypes with acceptable nutritional and quality characteristics. The identification of high-yielding broadly-adapted types requires multi-locational testing and selection. This is done in a wide range of environments with the cooperation of breeders from national programs. The flow of genetic material to national programs is shown in Figure 19.

A scheme of breeding for yield, as used at ICARDA, is shown in Figure 20. The use of an off-season growing season, screen houses, large isolation plots and the bagging of single plants are all very important in increasing the efficiency of the breeding program and in maintaining the genetic identity of advanced populations.

There is considerable emphasis in our breeding program on developing improved open-pollinated populations. However, synthetic varieties are becoming increasingly important, especially in northern Europe, and they offer a promising breeding method for faba beans. A synthetic variety is the population produced by inter-crossing, in a diallel, lines that have been selected for good general combining ability. A synthetic variety can be maintained by open

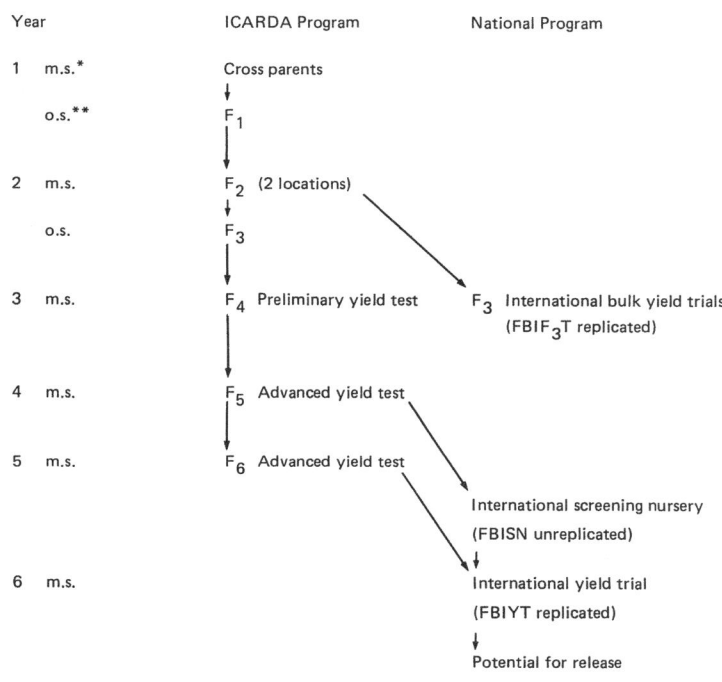

Year		ICARDA Program	National Program
1	m.s.*	Cross parents	
	o.s.**	F_1	
2	m.s.	F_2 (2 locations)	
	o.s.	F_3	
3	m.s.	F_4 Preliminary yield test	F_3 International bulk yield trials (FBIF$_3$T replicated)
4	m.s.	F_5 Advanced yield test	
5	m.s.	F_6 Advanced yield test	
			International screening nursery (FBISN unreplicated)
6	m.s.		International yield trial (FBIYT replicated)
			Potential for release

* main season
** off-season

Figure 19. Flow of genetic material to national programs.

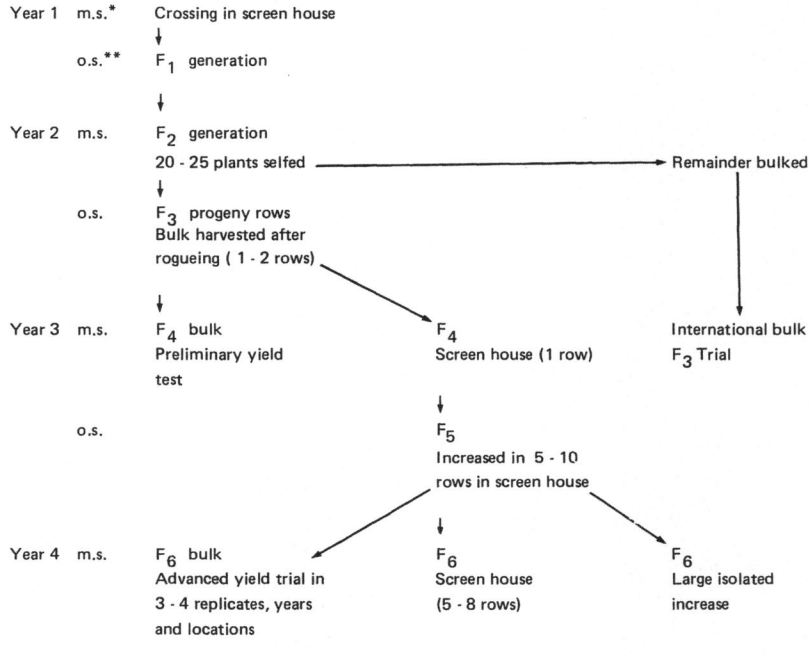

Figure 20. The scheme for breeding for yield at ICARDA.

pollination after its synthesis by hybridisation from the selected lines, or can be reconstituted from the original lines (Allard, 1960). Advantages offered by synthetics are:

i.   They utilize heterosis without requiring the regular controlled crossing needed for the production of F_1 hybrids.

ii.   Seed is cheaper to produce than for F_1 hybrids.

However, synthetic varieties are more expensive and time-consuming to breed, and their seed production is more expensive than in the case of open-pollinated populations.

Results

In replicated tests, 520 entries were evaluated at Tel Hadya, Syria, and 310 at Terbol, Lebanon. A seed yield of 6355 kg/ha was recorded, and several lines yielded more than the local check, which produced 4565 kg/ha. Seeds of most of these entries had been distributed to cooperators who grow the International Screening Nursery. The performance of some of the best entries from the BIYT-L (Bean International Yield Trial - Large Seeded) at Tel Hadya and Terbol for the two years 1979/81, is shown in Table 20. The entry, Elegant 5 MCI, yielded more than the check entry, Syrian local large (ILB 1814), across the locations for both seasons, except at AUB in 1980-81 where both entries gave the same yield. This demonstrates the wide adaptability of Elegant 5 MCI. Other entries had a higher yield potential, in some locations, as compared to the check, but were less widely adapted. The results highlight the difficulty of breeding for high yield and wide adaptation in varieties that are almost pure lines. Perhaps hybrids or synthetic varieties will display a wider adaptation.

Table 20. Seed yield, as a percentage of local check yield, of some selected genotypes in Syria (Tel Hadya (TH)) and Lebanon (AUB and Terbol (T)) in 1979-80 and 1980-81.

Entry	1979-80			1980-81			
	TH (Irrig.)	TH (Rainfed)	T	TH (Irrig.)	TH (Rainfed)	T	AUB
74TA 59	107	86	108	114	100	96	106
74TA 63	105	91	93	118	109	106	107
Reina Blanca	-	-	-	115	109	89	121
39 MB	113	101	105	111	103	87	80
Elegant 5 MCI	112	101	113	130*	102	102	100

* Significantly exceeding local check at 5% level.

ICARDA is the only research organisation that is trying to develop high yielding faba bean cultivars for low rainfall zones (300-350 mm). A summary of rainfed yield trials is shown in Table 21. These results indicate that yields significantly higher than the check were obtained only by the small seeded entries. Eight large seeded lines exceeded the check yield, although these increases were not statistically significant, and the heaviest yield was 4907 kg/ha. There is clearly a potential for developing suitable faba bean lines for low rainfall zones.

Table 21. Summary of rainfed faba bean yield trials at Tel Hadya (TH), Syria, and Terbol (T), Lebanon in the 1980-81 season.

Trial (location)	No. of test entries	Highest yield (kg/ha)	No. of entries exceeding check yield	No. of entries significantly exceeding check yield	Error coefficient of variation (%)
PYT-Small seed (TH)	110	3485	100	51	14.4-24.5
PYT-Small seed (T)	66	4173	44	4	15.4-17.4
BIYT-Large seed (TH)	15	4907	8	0	23.1
BIYT-Small seed (T)	23	3882	19	5	14.8
AYT-Large seed (TH)	19	3562	0	0	14.9
AYT-Large seed (T)	19	3327	13	0	10.5
AYT-Small seed (TH)	15	3101	9	6	18.6
AYT-Small seed (T)	15	3426	11	3	22.7

Breeding for disease and pest resistance

The major objectives of the ICARDA faba bean breeding program for disease and pest resistance are to identify sources of broad-spectrum, stable resistance to the major diseases and pests, and to incorporate this resistance into elite high-yielding genotypes. The program's activities started with a survey of diseases and pests, and an estimation of the economic losses that they cause. The survey has identified eight important fungal diseases: chocolate spot, ascochyta blight, rust, powdery mildew, root-rot/wilt complex, leaf spot, brown spot, and stem rot; four viruses: BBMV (broad bean mosaic virus), BYMV (bean yellow mosaic virus), BWV (bean wilt virus), and PMV (pea mosaic virus); seven important insect pests: stem borers, bruchids, aphids, leaf weevils, leaf miners, leaf hoppers and army worms; one predominant stem nematode; and two

species of broomrape, Orobanche crenata and O.aegyptiaca. The causal agents of
some important diseases and pests are listed in Table 22.

Table 22. Some important pests and diseases in faba bean, resistant lines and
their origin.

Disease	Causal agent	Resistant or less susceptible lines	Country of origin
Chocolate spot	Botrytis fabae	BPL 112,ILB 438, ILB 938.	UK, Ecuador, Ecuador.
Ascochyta blight	Ascochyta fabae	ILB 37; 80 Lat 14435-3, 14422-2, 14986-3, 14998-1.	Iraq, Lebanon, Lebanon, Greece, UK.
Rust	Uromyces fabae	80 Lat 15563-1,-2,-3, -4.	Canada.
Broomrape	Orobanche crenata	F402, BPL 561, 587.	Egypt, Ethiopia.
Aphids	Aphis fabae	Rustatt.	W. Germany.
Nematodes*	Ditylenchus dipsaci	80 S 51004-16, 51027-2, -5.	Aquadulce (Spain) x ILB 1560 (Sweden).
Bruchus*	Bruchus rufimanus	BPL 1025, 1085.	Afghanistan,Turkey.

* preliminary results

Reports of national programs in the region, and our own observations, have
indicated that both chocolate spot, (Botrytis fabae), and ascochyta blight,
(Ascochyta fabae), are the diseases with the greatest destructive potential to
the faba bean crop in the major producing countries in the region. These two
diseases have, therefore, been accorded the highest priority for resistance
breeding. Also, a very high priority has been given to developing Orobanche
resistant genotypes because Orobanche infestation occurs across a wide range of
environments in many countries, and it has the potential of devastating the
crop. Bruchid infestation in stored seeds reduces seed weight and quality, and
can lower their market value; because of its widespread distribution in many
countries in the region, more research is now centred on controlling bruchids.

Screening

Resistant lines and populations were identified by growing them at Lattakia,
Syria, where the environmental conditions are ideal for natural disease
development (Table 22). Sprinkler irrigation was used to provide conditions
suitable for epiphytotic disease development.

Chocolate spot

The materials tested included F_2 segregating populations and 526 BPL accessions. Of these, 14 were rated resistant and 48 moderately resistant. Approximately 50 F_2 individual disease-free plants were selected and will be evaluated as F_3 rows next year. In replicated trials, BPL 112 (U.K.), ILB 438 (Ecuador) and ILB 938 (Ecuador) were rated consistently resistant. The line BPL 938 was originally identified as resistant by the national program in Egypt, and it has shown a high level of resistance in the UK.

Ascochyta blight

The material tested included BPL accessions and F_2 segregating populations. Of over 500 BPL accessions, 49 were promising, and eight of them were highly resistant. Over 1000 single individual resistant and moderately resistant plants were selected for progeny row evaluation. The lines listed in Table 22 showed a high degree of resistance across replications. A faba bean host differential set was developed on the basis of disease reaction and the number of pycnidia per unit area (total of nine lines: two highly resistant, one resistant, two moderately resistant, two susceptible, and two highly susceptible). The differential set was distributed to a few national and international programs to study physiological races in Ascochyta fabae.

Rust

The lines 80 Lat 15563-1, -2, and -3 were originally identified as being resistant in Manitoba, Canada, and were also found to be resistant in Syria.

Orobanche

Most screening for resistance was carried out at Kafr Antoon, Syria, where the level of soil infestation of O. crenata is very high. The materials tested were BPL accessions, segregating populations, and the International Orobanche Nursery (ION). Forty three out of 420 BPL accessions were identified to have a lower level of infestation than Family 402 (resistant check). The 28 BPL lines that were identified in 1979-80 have maintained their performance in 1980-81. A total of 2800 F_2 individual plants and 31 F_3 progeny rows were better than the local susceptible check. Two years data from the ION in Syria indicated that BPL 561, 587, and 811 had lower Orobanche infestation as compared to the resistant check, Family 402.

Bruchids, Aphids, and Nematodes

Preliminary results indicated that the lines listed in Table 22 may be resistant, but further evaluation is needed.

Information on the biology and epidemiology of the diseases is being sought where essential basic information is lacking. Crosses among resistant and resistant x susceptible genotypes are being made for each disease to try to detect the number of resistant genes available. This will allow the development of cultivars with broad-spectrum, stable resistance for west Asia and the development of genetic stocks for each disease, which can then be used by national programs to counter their disease problems (Figure 21).

Year 1 m.s.*   Crossing in screen house
                      ↓
        o.s.** F_1 — off-season
                      ↓
Year 2 m.s. F_2 — Single plant selection and selfing of resistant
                              plants under disease pressure.
                      ↓
        o.s. F_3 — Bulk harvest of progeny rows from selected
                      ↓ F_2 plants.

Year 3 m.s. F_4 — Single plant selection from bulks, and selfing of
                      ↓   resistant plants under disease pressure.

        o.s. F_5 — Seed increase for resistant plants as progeny rows.
                      ↓
                  F_6 — Genetic stocks.

*   main season
**   off-season

Figure 21.   The scheme for breeding for resistance to diseases and/or pests at ICARDA .

Breeding for quality and other traits

Germplasm is being evaluated for the following characteristics: protein content, anti-nutritional factors that cause favism, the correlation between dry seed yield and green bean yields, autofertility, male sterility, plant ideotype (determinate growth habit), and drought tolerance.

Protein content

BPL numbers 171, 189, 296, 303, 355, 405, 494, 505, 520 and others have been identified from the world collection as being high in protein (30 to 35%). Unfortunately, most of these lines are unadapted to conditions in western Asia, so several crosses are being made to improve their genetic background.

Anti-nutritional factors

A high frequency of favism (a human disease that can be fatal) has been reported from several Middle Eastern countries. It is mainly associated with a high consumption of faba bean containing high levels of vicine and convicine in the cotyledon.   ICARDA, in collaboration with the University of Manitoba, has initiated a project to survey favism, and to screen and identify genotypes that are low in vicine and convicine. The cultivar, Triple White, was identified as being low in both substances as well as having a white hilum. This association, as well as the mode of inheritance, will be further investigated.

Green bean production

Hawtin and Stewart (1979) have estimated that about two-thirds of the large seeded faba bean (V. faba major) crop in the ICARDA region is harvested for use as a green vegetable, while one third is used as a dry seed. Studies conducted at Tel Hadya, using a wide range of different types of materials, indicated that dry seed and green pod production are highly correlated (Table 23). Hence, a decision has been made to discontinue the green bean production project, since selecting for one trait is likely to ensure the presence of the other.

Table 23. Correlations between dry seed yield and green bean yields in the 1980-81 season at Tel Hadya.

Trial	No. of entries	Correlation coefficients between dry seed yield and green pod yield with:		
		One picking	Two pickings	Three pickings
BIWT-L-81	10	0.41	0.72*	0.88**
BPL-collection	56	0.42**	0.43**	0.48**
ILB-collection	25	0.01	-	-0.09

* P <0.05
** P <0.05 >0.01

Autofertility

A total of eight genotypes, four from the Middle East and the others from Europe, were evaluated for autofertility under the screen house. This was done by comparing the number of pods per plant and seed yield per plant with either tripped or non-tripped flowers. Cultivars originating from the ICARDA region possess higher autofertility than cultivars originating from northern Europe (Table 24).   Bagging individual plants in our segregating populations for subsequent progeny row evaluation will automatically ensure that autofertility will remain at an acceptable level (especially in crosses made with the European types).

Table 24. Effect of tripping on the pod and seed yield of some selected genotypes of faba beans grown in cages to exclude pollinating insects.

Cultivars	Pods per plant		Seed yield(g) per plant	
	Tripped	Untripped	Tripped	Untripped
Syrian Local Small	6.4	8.9	4.8	7.6
Giza-3 (Egypt)	9.0	8.8	10.4	12.0
Giza-4 (Egypt)	7.8	8.8	10.3	11.0
Hudeiba-72 (Sudan)	5.6	7.1	4.9	6.6
Seville Giant (Spain)	3.5	3.8	17.7	16.8
Express (Spanish-type)	5.1	5.0	16.1	9.7
Throws MS (UK)	3.6	0.8	3.7	0.5
Maris Bead (UK)	3.2	1.9	2.6	0.9

Male sterility

Existing male sterile lines are limited in number, and are in a genetic background which is not adapted to the Mediterranean region. Several back-crosses are being made to improve the performance of these lines.

Plant ideotype

The investigation of plant ideotype is a long term project, where the breeder and physiologist have to identify, develop and evaluate different plant ideotypes such as determinate vs. indeterminate growth habit, tall vs. short, low vs. high number of branches, and others. Their evaluation will be conducted in a wide range of environments, and the contrasting traits will be compared using genotypes with a similar genetic background.

Several advanced breeding lines from the determinate types have been developed and will be compared with indeterminate types for their yielding ability next year.

Interspecific crossing in Vicia

ICARDA, in cooperation with the University of Reading, have recently initiated a project to identify the genetic and non-genetic factors that are limiting the transfer of useful traits, especially disease resistance, from the wild species to the cultivated species V. faba. Attempts will be made to cross V. faba with V. narbonensis, V. bithynica, and V. hyaescyamus.

References

Allard, R.W. 1960. Principles of plant breeding. New York, U.S.A.: John Wiley.

FAO. 1980. FAO Food Production Yearbook. Rome, Italy: Food and Agriculture Organization of the United Nations.

Hawtin, G.C. and Stewart, R. 1979. The development, production and problems of *Vicia faba* in West Asia and North Africa. FABIS 1:7-9.

16. TAXONOMY, DISTRIBUTION AND EVOLUTION
OF THE LENTIL AND ITS
WILD RELATIVES

J.I. Cubero

Department of Genetics, Escuela Tecnica Superior
de Ingenieros Agronomos, Cordoba, Spain

Taxonomy of Lens Miller

There are five species in the genus Lens Miller, namely Lens montbretii (Fisch. & Mey.) Davis & Plitmann, L. ervoides (Brign.) Grande, L. nigricans (M.Bieb.) Godron, L. orientalis (Boiss.) Handel-Mazzeti, and L. culinaris Medikus.

L. montbretii is well separated from the other four species, on the basis of strong pubescency, and its long leaves with many leaflets and long, branched tendrils. Its peduncle is much shorter than the rachis, and it has long and elongated pubescent pods. It is a species of doubtful systematic position and is therefore not included in the following discussion.

The four other species constitute the culinari group. Table 25 summarises the main differential characters in this group. It can be used as a 'key', perhaps with more accuracy than a typical dichotomous one. The stipule morphology is very variable both between and within culinaris cultivars. It is also possible to find strong pubescence in both orientalis and culinaris. Intermediate forms are not infrequent. In fact, most, if not all, the characters shown in Table 25 would have to be preceded by the word 'generally'. The floral morphology of this group of four species is shown schematically in Figure 22.

The position of L. orientalis and L. nigricans

Commenting on the possibility, suggested by Davis and Plitmann (1970), of nigricans being cultivated in some areas of Turkey, Ladizinsky (1979 a) questions if they are not, in fact, culinaris forms with nigricans stipules. He also points out that the populations of nigricans reported from typical orientalis areas can be, in fact, orientalis populations with nigricans stipules. If the morphological differences among these species are such that only one character can lead us to a wrong taxonomic decision, what is then the real difference between them? For example, how can we know that we are working with nigricans and not culinaris with toothed stipules, or with orientalis and not ervoides with awned peduncles? Other examples can be added. In Boissier's description of Ervum cyaneum (=L. orientalis) the calyx teeth are only twice the

J.R. Witcombe and W. Erskine (eds.) Genetic Resources and Their Exploitation - Chickpeas, Faba beans and Lentils. ISBN-13:978-94-009-6133-3 (PB)
©1984, Martinus Nijhoff/Dr W. Junk Publishers for ICARDA and IBPGR.

Table 25. Differential characters of the <u>culinaris</u> group.

Character	ervoides	nigricans	orientalis	culinaris
1 Toothed stipules	no	yes	no	no
2 No. leaflets/leaf	4-8	4-10	6-12	(4)*10-20
3 Rachis length (mm)	< 10	±15	±20	40-50
4 Aristate peduncle or awn	no	yes	yes/no	yes
5 Calyx teeth/corolla length ratio	<<1	>=1	<=1	<=>1
6 Peduncle/rachis length ratio	> 1	> 1	> 1	<=1
7 Pod pubescence	yes	no	no	no **
8 Max. diam. seed (mm)	3	3.5	3.5	9 ***

*   Lowest value recorded, 10-20 normal range.

**   Generally

*** Some forms with diam. 3 mm

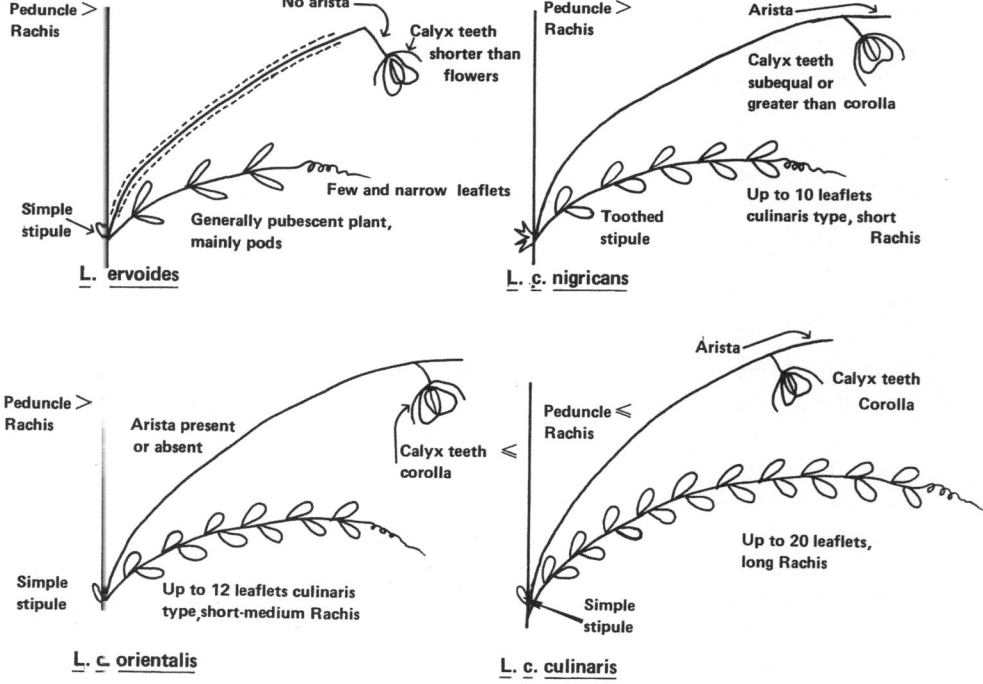

Figure 22. A diagram of the floral morphology of the <u>culinaris</u> group of the genus <u>Lens</u>.

length of the tube and clearly shorter than the corolla, and its peduncle (aristated) is much longer than the rachis: a strange mixture of ervoides, nigricans and orientalis characters, complicated by the fact, reported by Boissier, that 'intermediate forms (with orientalis) are found'. Furthermore, it is not rare to find herbarium specimens that have been successively classified as nigricans and orientalis and that show characters from both. There are orientalis without awned peduncles, and ervoides with them. There are culinaris and orientalis from Syria and Turkey which are strongly pubescent.

Fortunately, two groups have initiated a different approach, providing evidence that can be used in a taxonomy based on the biological concept of the species.

First, Williams et al. (1974, 1975) compared nigricans, orientalis and culinaris (ervoides was not included in the study) using principal component analysis, and found culinaris and orientalis closer to each other than either of them to nigricans. As the pollen exine morphology was also different for nigricans when compared with the two other species, they suggested that orientalis should be a subspecies of culinaris; all three forming a compilo-species. Later, they considered nigricans also as a subspecies of culinaris.

Second, Ladizinsky (1979 a) made crosses between culinaris (three accessions, having extreme phenotypes for certain characteristics), orientalis (four) and nigricans (one):

i. Culinaris x orientalis hybrids and the F_2 generation grew normally. Meiosis was almost normal, with the presence of a quadrivalent in most of the cells examined and, in very few cases, a trivalent and one univalent. This result supports the proposal of Williams et al. of considering orientalis as a subspecies of culinaris. The similarity between them, and even the possible relationship of common descent, had been pointed out by Barulina (1930) and by Zohary (1972), but they did not mention possible con-specificity.

ii. Culinaris x nigricans hybrids also developed normally. Tendrils were more developed than in either of the parental species, and many stipules were semi-hastate (a nigricans character but also present in culinaris) and eventually dentate (pure nigricans, excepting the data in Ladizinsky's work). Meiosis was more irregular than in the previous case. Ladizinsky concluded that these species differ by three interchanges instead of the one suggested in the culinaris x orientalis case. He notes that, in spite of this meiotic irregularity, the fertility of the hybrids was rather high, producing at least one seed/pod. However, pollen fertility and the percentage of flowers developing was lower than that of the parental species. Ladizinsky's evidence strongly supports the conclusions of Williams et al. (1975). It seems that there is only a slight difficulty in genetic flow between nigricans and culinaris. If this flow is not intense in nature, it is because of strong autogamy in the cultivated species, and probably also in the wild ones. The data available to us, both on morphological and genetic grounds, suggest that nigricans can be considered more a subspecies of culinaris than as a separate biological species.

The only disagreement with the view that L. orientalis and L. nigricans are subspecies of L. culinaris comes from karyological studies. According to Ladizinsky, culinaris and orientalis have the same karyotype. Nigricans has a slightly different karyotype, with two metacentrics instead of the two submetacentrics of both culinaris and orientalis. If the difference is due to translocations there is no reason to give a specific rank to nigricans. But this is not all: other authors have found karyotypes different to those described by Ladizinsky. In particular, Sindhu (pers. comm.) has found that the main difference between the karyotype of nigricans and those of culinaris/orientalis was that the larger chromosome was submetacentric in the latter, and metacentric in nigricans.

Strains of culinaris spp. microsperma differ in the number of metacentrics, submetacentrics and acrocentrics, and that some cultivars of culinaris spp. macrosperma do not have the usual secondary constriction in one of the pairs of metacentrics. So, then, again the variation found in culinaris overlaps with nigricans. The most logical status for nigricans seems to be that of a subspecies. The meiotic studies on the hybrids obtained by Sindhu and Slinkard in their extensive program of crosses between culinaris, nigricans and orientalis (Sindhu, pers. comn.) will clarify this problem.

The position of L. ervoides and L. montbretii

Up to now, crosses between culinaris and ervoides have produced non-viable seeds (Ladizinsky, 1979 b). So, even when it is possible to find intermediate forms between ervoides and, at least, culinaris, it is necessary to keep the specific status for this taxon. Many more crosses are needed, not only in this case, but also for orientalis x nigricans, where the resultant albino hybrids die shortly after germination.

The position of montbretii in regard to the culinaris group is clear from morphology, but nothing is known from the genetical viewpoint. L. montbretii and L. lunata seem to be intermediate forms between Lens and Vicia, as is also the section lenticula of Vicia, and the study of all of these forms is interesting, not only on theoretical, but also on practical grounds.

Summary of the taxonomy of Lens

i.   L. orientalis and L. nigricans belong to the primary gene pool of L. culinaris. They seem to be con-specific.

ii. Awaiting new data, L. ervoides has to be kept in its specific status, but it could belong to the secondary gene pool of the cultivated species.

iii. L. montbretii looks very different from all of the culinaris group; it is unlikely to be removed from its specific status.

Distribution

The geographical distribution of these five species has been derived mainly from the Kew collection, but also from Floras of the region where Lens is known to grow naturally (Figure 23). For convenience, Uganda has not been represented on the map; only L. ervoides is known to grow wild there.

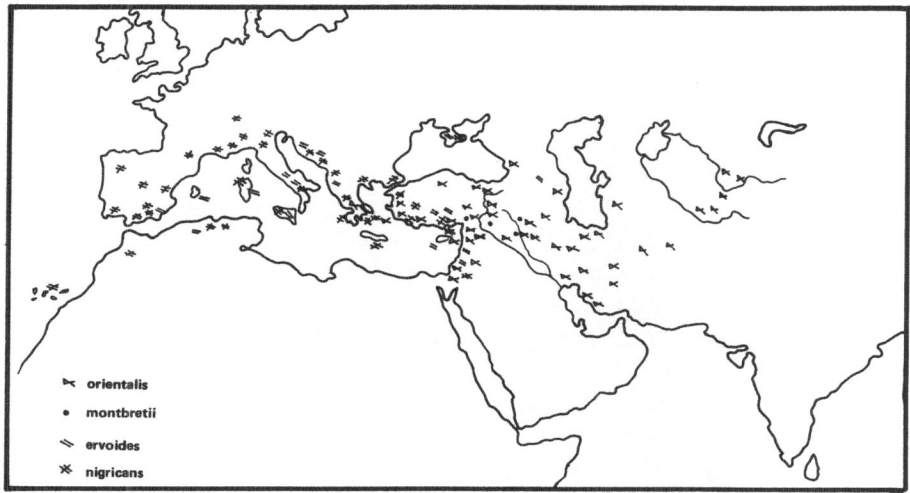

Figure 23. Distribution of wild lentils.

Taxonomy of the cultivated lentil

The first person to study the taxonomy of the cultivated lentil was Alefeld (1866), who recognised eight subspecies, including as such both orientalis and nigricans (the specific name adopted by Alefeld was L. esculenta Moench).

ssp. schniffspahni (ex L. sch. indica Alef.) (= orientalis)

ssp. himalayensis (ex E. himalayense Al. Braun & Bouché) (= nigricans)

ssp. punctata (ex E. punctatum Alef.) (= culinaris)

ssp. hypochloris

ssp. nigra (ex E. nigrum and E. camelarum) (= culinaris)

ssp. vulgaris

ssp. nummularia

ssp. abyssinica

There are too many subspecies, even for an old-fashioned taxonomist. However, one may be grateful to Alefeld because he reduced the number of species, including in esculenta (=culinaris) even nigricans and orientalis. His subspecific names were taken later by Barulina (1930) to name varieties.

The most detailed and complete study of the cultivated lentil was made by Barulina (1930), a disciple of Vavilov. Studying thousands of accessions from all parts of the world, she recognised two subspecies, according to the size of the seed. The names of the subspecies were fortunately chosen, not only because many other characteristics are correlated with the seed size, but because the seed was the main objective of man's selection, and the evolutionary process of this species as a crop has been dependent on this fact.

In contrast to Alefeld, Barulina subdivided the subspecies using characters which were not very sensitive to environmental fluctuations. She also considered the geographical distribution of clusters of characters, so defining regional groups or grex. In her study to define subspecies, the main characters were pods, seeds, and distinct differences in the length of flowers. Secondary characters included size of leaflets, length of vegetation, and height of plants. Important characters with which to define groups within the subspecies included dehiscence, length of calyx teeth, and number of flowers per peduncle. Barulina labelled simple allelic differences with Latin; in the description of subspecies and geographical groups the Latin names will be retained for both kinds of taxa. Even though Harlan and de Wet (1971) suggest that Latin names are used only for subspecies at the infraspecific levels, in Barulina's work a Latin name for a grex is simple and descriptive.

Barulina's description follows below. The geographical distribution of subspecies and grex are shown, according to Barulina's own map, in Figure 24.

i.   ssp. macrosperma (Baumg. pro var.) Barulina
     Large pods (15-20 x 7.5-10.5 mm) generally flat. Large flattened seeds (6-8 mm diam.). Cotyledons yellow or orange. Flowers large (7-8 mm long), white with veins, rarely light blue. Peduncles with 2-3 flowers. Calyx teeth long. Large leaflets (15-27 x 4-10 mm), oval, length to width ratio = 3-3.5. Height of plant 25-75 cm. No geographical groups. 12 varieties.

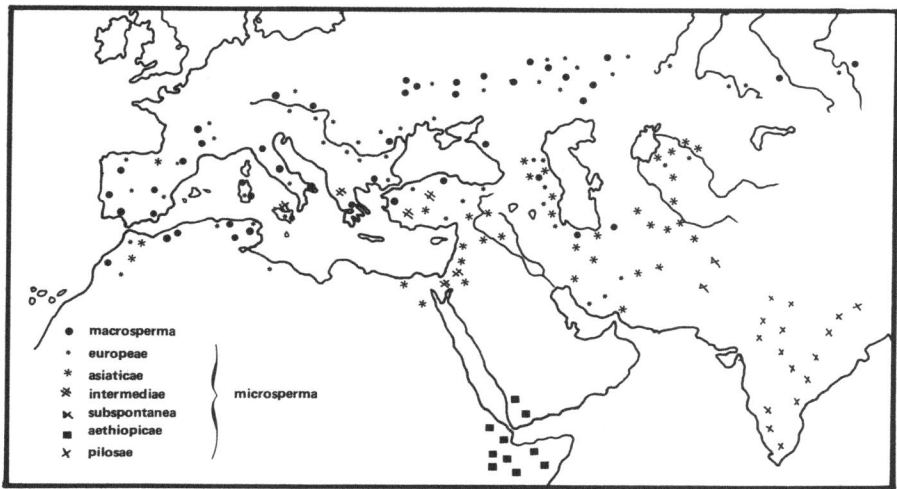

Figure 24. Distribution of cultivated lentils (Barulina, 1930).

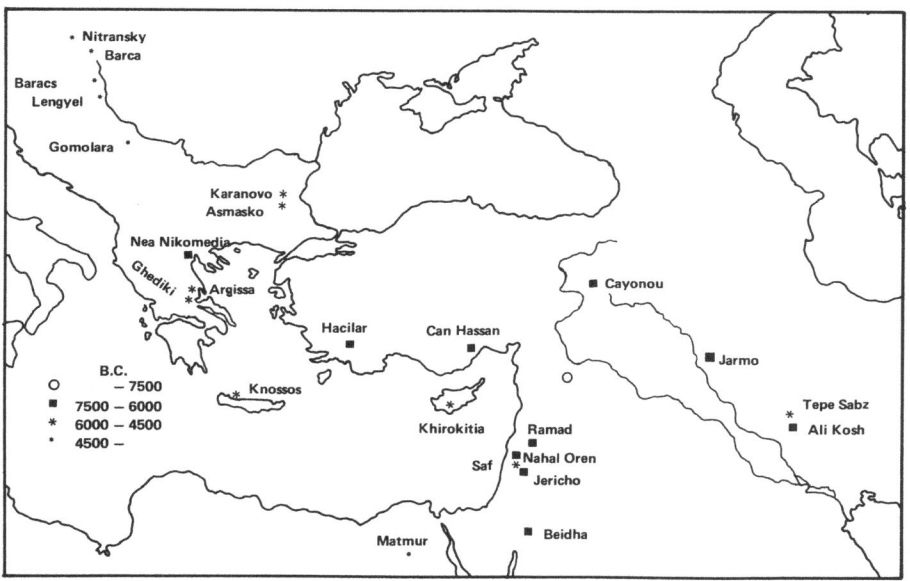

Figure 25. Distribution of archeological remains of lentils.

ii.   ssp. microsperma (Baumg. pro var.) Barulina

Pods small or medium (6-15 x 3.5-7 mm), convex. Seed flattened-subglobose (diam. to thickness ratio = 1.5-3), small or medium (3-6 mm diam.), diverse in colour and pattern. Flowers small (5-7 mm long), from white to violet. Peduncles 1-4 flowers. Leaflets small (8-15 x 2-5 mm) elongated linear or lanceolate (length to width ratio = 4-5). Height of plant 15-35 cm.

There are six geographical groups (grex) and 46 varieties of microsperma:

i.   Grex europeae. 2-4 flowers (3) per peduncle. Seeds 4-5 mm diam. Flowers white with veins. Calyx teeth much longer than corolla. Leaflets medium sized. Mid-season forms. Seeds diverse in colour. Six varieties.

ii.   Grex asiaticae. 1-3 flowers (2) per peduncle, blue or light blue. Seeds 3-5 mm diam. Calyx teeth equal or rarely longer than corolla. Leaflets small, narrow, elongated. Seed diverse in colours. Early forms. 22 varieties.

iii. Grex intermediae. 1-3 flowers, white with veins. Seeds 5-6 mm diam. Calyx teeth longer or equal to corolla. Leaflets small or medium. Nine varieties.

iv.   Grex subspontaneae. Pods dehiscent, purple coloured before maturity, brown or black at maturity. Seeds small (3-3.5 mm diam.) black or blackish. 1-2 flowers per raceme, violet-blue. Calyx teeth much shorter than corolla. Leaves with 3-5 pairs of small leaflets (8.5-9 x 2.2-2.8 mm). Plant dwarf. Two varieties.

v.   Grex aethiopicae. 1-2 flowers per peduncle, violet-blue. Calyx teeth much shorter than corolla. Leaves 3-7 pairs of elongated leaflets (13-15 x 2.9-3.7 mm). Pods with elongated apex. Seeds brown with black dots or black. Two varieties.

vi.   Grex pilosae. 1-2 flowers (2) per peduncle. Seed reddish with black dots or black. Calyx teeth much shorter than corolla. Tendrils short, sometimes rudimentary. Leaves with 3-6 pairs of leaflets. Plant dwarf, pubescent. Five varieties.

Barulina's work has not been developed by others. It can be completed by studies on chemosystematics, chromosome banding, and the relationship of culinaris with other Lens species. It is not difficult to reconcile Barulina's system and that proposed in the taxonomy of Lens Miller for the genus. Both Barulina's subspecies (macrosperma and microsperma) have to be considered races of ssp. culinaris. I think that it is useful to keep these Latin names, as well as those of the grex, because they are concise and descriptive, even without crediting them to any author. In short:

Lens culinaris Med.

ssp. nigricans (M. Bieb.) Cubero

ssp. <u>orientalis</u> (Hand.-Maz.) Will. Sanch. et Jack.
ssp. <u>culinaris</u> (Briq.) Will. Sanch. et Jack.
    race <u>macrosperma</u>
    race <u>microsperma</u>
        grex <u>europeae</u>
        grex <u>asiaticae</u>
        grex <u>intermediae</u>
        grex <u>subspontaneae</u>
        grex <u>aethiopicae</u>
        grex <u>pilosae</u>

Recent expeditions to collect germplasm have been able to recover most of the forms described by Barulina except the grex <u>subspontaneae</u>.

Origin, domestication and evolution of cultivated lentils

The centre of origin of the crop in the Fertile Crescent

Many places of archaeological interest to lentils are known (Figure 25). The evidence suggests that lentils were first cultivated in the 'Fertile Crescent' and southern Turkey, spreading later towards the Nile, Greece and central Europe following the Danube. Of course, the extent of our knowledge of the different regions of the world is unequal, but up to now the new findings do not modify the pattern discernible in (Figure 25 above). In fact, the pattern follows the spreading of the neolithic agricultural techniques.

Even though Van Zeist (1976) has pointed out the uncertainty of dating archaeological sites, it seems obvious that the oldest findings, to date, of lentils (as well as wheat and some other crops) are in the Near East, from the viewpoint of both absolute dates and cultural background.

The absence of lentil remains in eastern and southern Spanish sites dating back to 4700 B.C. is surprising. Hopf (1971) has studied materials from sites dating from 4700-4000 B.C. (four sites) and 3000-1500 B.C. (three). In all seven she found abundant remains of wheat, pistachio nut, and in one case <u>Vicia faba</u>, but she found no remains of either cultivated or wild lentils.

More information is needed from the western Mediterranean countries. It is not easily credible that the El Argar culture of S.E. Spain, which was probably the origin of the 'bell beaker' culture (c. 2000 B.C.), was not familiar with lentils, and also that farmers who sowed wheat, barley and even faba beans neglected lentils.

Thus, it seems that the place of origin of lentils as a crop is the Near East around the Fertile Crescent. Recently, Ladizinsky (1979a) has claimed that 'it is apparent that lentil was not domesticated at the southwestern part of the Fertile Crescent in the Middle East' because the <u>orientalis</u> strain used by him differed from the cultivated one by a reciprocal translocation. But it has been shown above that chromosomic polymorphism is found even within cultivated strairs. It is difficult to place the exact location where lentils were first domesticated; that is, it may never be known whether or not the southwestern part of the Fertile Crescent was involved.

Afghanistan as a centre of origin

Barulina (1930) suggested the eastern border of southwestern Asia as a possible 'centre of origin' of the cultivated lentil. She based her suggestion on the fact that the region between Afghanistan, India and Turkestan, that is, the Himalaya-Hindu Kush junction, showed the highest proportion of endemic varieties of the cultivated species: all kinds of rare forms, like subspontaneae, were concentrated there, showing great morphological and physiological variation. However, she noticed that the area of distribution of wild lentils did not overlap well with that of the domesticated ones, except that the eastern part of the orientalis area of distribution also reaches Turkmenia.

To discuss this problem, let us first compare the distribution of wild and cultivated lentils (Figures 23 and 24) bearing in mind the description of Barulina's grex of L. culinaris. It can be observed that three of the microsperma groups (grex) are restricted to very specific areas: pilosae to the Indian subcontinent, aethiopicae to Ethiopia and Yemen, and subspontaneae to the Afghan regions closest to India. These three groups are characterized by very small and dark coloured seeds, violet flowers, few flowers per peduncle, calyx teeth much shorter than the corolla, few leaflets per leaf, and dwarf plants. Against this common background, each one of these grex shows a distinct character: grex pilosae a strong pubescency, grex subspontaneae very dehiscent pods, being purple coloured before maturity, and grex aethiopicae pods with a characteristic elongated apex. So, each one of the three isolated groups seems to show some distinct fixed character and a cluster of primitive characteristics closely related to L. orientalis. A comparison between Figures 23 and 24 leads to the conclusion that the distribution of grex subspontaneae only overlaps with L. orientalis, and grex aethiopicae only with L. ervoides, and grex pilosae does not overlap with any wild lentil.

In contrast, the other three microsperma grex (europeae, asiaticae and intermediae) and macrosperma are rather cosmopolitan and are clearly intermixed. Seeds are variable in size, but in general they are wider than 4 mm. They have more flowers per peduncle and more leaflets per leaf, and the calyx teeth are equal to or greater than the corolla. White flowers are common, and seeds are diverse in colour. These microsperma grex, as well as macrosperma, overlap, to a greater or lesser extent, with all the known wild lentils.

When the above pattern of distribution of the wild and cultivated forms is considered, the great number of endemic forms found in the Afghanistan - India - Turkmenian border region cannot be explained by the hypothesis of this area being the 'centre of origin' of the cultivated species. It is difficult to explain how culinaris is much richer there, from a genetic point of view, than orientalis. It is also difficult to explain the coexistence of grex aethiopicae and L. ervoides in Ethiopia and, finally, it is no easier to explain the pubescence characterizing grex pilosae in India, pubescent forms of both L. orientalis and L. culinaris being present in Palestine, Syria and Turkey.

The great diversity of forms found by Barulina in the Hindu Kush-Himalaya border is best explained by the idea of a microcentre, that is, a region where a great variation can be found. In this case it is probably correlated with a mountainous topography of many isolated valleys. Such isolation allows genetic drift with drastic genetic fixation and loss to occur. Supporting the idea that Barulina's proposed region is not the 'centre of origin' but an important secondary centre of diversity, is the fact that no lentils have been found in sites dating back to the fifth millenium B.C. in Turkmenia (Djebel, Djeitum).

Perhaps future findings will modify our present schemes, but the known data point to southeastern Turkey and northern Iraq as the place of domestication of lentils, that is, the 'centre of origin' of the crop. There, some populations of what we now call orientalis, perhaps mixed with nigricans, were unconsciously subjected to selection, and, little by little, distinct cultivated populations were built up to give culinaris, a new crop.

The spreading of lentil culture

Assuming that the domesticated forms arose in southern Turkey - northern Iraq, how did the culture spread to other regions? Figure 25 shows some possible ways. Greece was reached very soon: Renfrew (1969) suggested that some of the Greek lentils were nigricans, and she thought this form was the ancestor of the cultivated ones. More archaeological studies are necessary in Thrace. Probably it had a similar cultural background to that of Anatolia (particularly to that of Lidia, Misia and Paphlagonia): obviously connected to the 'nuclear area' of the Near East.

From the neolithic culture of Thrace lentils found their way to central Europe via the Danube. We do not know how lentils reached Russia, but Figure 25 shows that central Russia, and then Siberia, was more likely reached from the western coast of the Black Sea, or from any site in the Danube Valley, than from the Transcaucasian-Caspian or from the Turkmenian regions. The varietal pattern seems to be very clear.

The Nile Delta was probably reached very soon, but its soil is not the best one to preserve carbonised organic matter for thousands of years. However, the arrival of lentils must have been very early for two main reasons:

i.   The Nile Delta was a cultural unity with the Near East.

ii.   The grex _aethiopicae_ exhibits very primitive characters, meaning that lentils probably went through the Nile valley at a very early stage of domestication.

Lentils did not reach India before 2000 B.C., but again there is the problem of availability of archaeological findings. This date can be certain for continental India, but it is well known that Mohenjo-Daro had contact with Sumerians and Akkadians in Mesopotamia. So, at least for the Indus valley, that date will probably be pushed back one more millenium at least.   It is interesting to note that, on linguistic grounds, De Candolle (1882) wrote, 'It may be supposed that the lentil was not known in this country (India) before the invasion of the Sanskrit-speaking race'. This invasion is supposed to have happened before 2000 B.C.

Evolution of the cultivated forms

Earlier suggestions of _L_. _orientalis_ being the material out of which the cultivated forms developed were confirmed by modern studies, and _nigricans_ could also be present in the birth of the _culinaris_ forms. Let us first establish the main features which differentiate the domesticated from the wild forms. In general, these are a greater height, a longer rachis and a greater number of leaflets per leaf, a greater area of leaflets, a greater number of flowers, and larger pods and seeds. It is easily seen that all these characters are related to the increase in yield required in any crop. Even the characteristic feature of shorter peduncles than rachis, differentiating ssp. _culinaris_ from all the other forms of the _culinaris_ group, is explained in such a way; high yield is aided by an increase in rachis length but not by an increase in peduncle length.

We have seen that there are cultivated groups with primitive characters, and others which are clearly evolved. We also know that in historical times, there were 'blackish' and 'not sweet' lentils, 'rounder' and 'normal' in shape, 'wild' varieties and so on. Little by little, farmers were selecting better varieties, with higher yields and better cooking qualities. We have seen that _orientalis_, _nigricans_ and also probably _ervoides_ were the sources of variability that was released by inter-crossing.   When domesticated lentils were spreading in all directions, they crossed with their wild relatives which contributed new genes from time to time. The wild relatives may well have been 'companion weeds' that were carried with the crop.

Was the distribution of the wild lentils the same as it is nowadays? Figure 23 shows a very distinct pattern of distribution: _orientalis_ is the only one spreading eastwards, _ervoides_ to the south (Ethiopia and Uganda), and both _nigricans_ and _ervoides_ to the west. Now, if the distribution of wild lentils were the same as at the present time before the existence of domesticated forms, it should be possible to find regional ecotypes from Spain to eastern Turkey of L. _nigricans_ and L. _ervoides_, and from Turkey to Turkmenia of L. _orientalis_. Perhaps the study of living collections will widen our knowledge but, at least working with herbarium specimens, it is impossible to discover any such regional differences. Even the easternmost _orientalis_ are very similar to the westernmost _nigricans_. The few varieties described are the product of taxonomists' imagination, rather than real entities.

It would be useful to test if there are any differences between the eastern and western populations within _nigricans_, _orientalis_ and _ervoides_.

The following new hypothesis is proposed. Lentils were domesticated, as explained above, in the foothills of the mountains of southeastern Turkey and northern Iraq. The raw materials were populations of _orientalis_, but primitive farmers would also have experimented with _nigricans_ and _ervoides_ populations or, much more logically, with mixed populations of all three forms, even when the relative proportions could be very different from one place to another. Variability for palatability, digestibility or toxicity would have produced an automatic selection towards convenient characteristics.

Meanwhile, these mixed populations (more or less purified) with new and useful characters, migrated east, west and southwards. If there were no such wild forms in their actual distribution areas before domestication, then these forms were authentic companion weeds of the domesticate. In fact, collections of wild lentils in recent times have been always carried out in disturbed habitats, a good demonstration of their weedy nature. The spreading of mixed populations could release wild materials up to the moment when all the populations were formed of pure domesticated plants. There are no wild lentils in central Europe, Russia or Siberia, or even in India, where the cultivated lentil is extremely primitive and has migrated the furthest.

The present distribution of the wild species, if we accept the hypothesis that it is a result of migration, can be explained in either of two ways:

i. The wild species spread from mixed populations that were richer in _nigricans_ and _ervoides_ in the western fringe, and in _orientalis_ in the eastern fringe of the original area (S. Turkey - N. Iraq).

ii. _Orientalis_, _nigricans_ and _ervoides_ may have different ecological requirements, for example for photoperiod sensitivity. As the wild species accompanied the crop in its outward migration so _nigricans_ and _ervoides_ survived in the north and east and _orientalis_ in the west.

It should be noted that the two possibilities outlined above are not exclusive and both of them may be true.

Whether the wild lentils were already distributed in their present zones or whether they went along with the domesticates, the wild material seems to have given to the more evolved domesticate different characters in the west from those in the east. Comparing again Figures 23 and 24, and bearing in mind the description of regional groups, we can see that the 'western wing' is characterized by big seeds, a great number of leaflets, and calyx teeth longer than corolla. The opposite is true for the 'eastern wing', where seed size (and all the other characters) is less, even with the amazing genetic variability recorded by Barulina. It seems that, in the east, genetic variation is greater for qualitative, rather than quantitative, characters.

The easternmost and southernmost populations have been already discussed: they look like primitive materials that have fixed some characters. It is tempting to conclude that the original stock of pilosae was in southern Turkey or northern Syria where hairy L. orientalis and L. culinaris forms are found. Pilosae, without any companion weed, had very few possibilities to evolve, even in such a huge region. A similar case seems to be that of aethiopicae. Perhaps, L. ervoides is really more a 'weed' than a 'companion weed', unless it is responsible for the short calyx of aethiopicae forms.

Has L. nigricans contributed essentially different genetic material than orientalis to the crop? The data reported above seem to suggest so, but this needs practical confirmation. Nigricans could have provided the genetic raw material for bigger seeds and all its correlated characters. Also, it could have been the source of the genes for adaptation to the Mediterranean-European environments, if it is positively demonstrated that orientalis lacks variability for adaptation to such environments. Remember also that crosses between culinaris and nigricans produced a greater development of tendrils and that culinaris shows, even rarely, branched tendrils. Calyx teeth are also greater in the 'western wing'; this can be correlated with leaflet area, but it is also typical of nigricans.

Of course, there is always an alternative possibility to explain the origin of large seeded varieties: Phoenicians and Greek traders, very active during the first millenium B.C., would have exchanged eastern and western materials. But again, why do large seeded varieties only appear in the western wing?.

The centre of origin of genus Lens

The area from western Turkey to northern Iraq (Kurdistan) contains not only all the wild lentils, including montbretii, (a doubtful lentil but undoubtedly related to Lens), but also Vicia lunata, a no less doubtful species that can be defined as a 'lentoid' Vicia.

We can suppose that this region is the cradle of 'lentoid' characters such as flattened pods and seeds. It could be a serious candidate to be the authentic centre of origin of Lens.

Summing up:

i. Lentils were domesticated in the area between southwestern Turkey and Turkestan from orientalis populations where nigricans was present. Culinaris was probably built up from orientalis populations with enough variability to colonise Mediterranean, Ethiopian and Indian environments or, alternatively, from intermediate nigricans-orientalis forms and/or from mixed populations of these.

ii. Nigricans and orientalis were companion weeds of culinaris when the crop moved west and eastwards, respectively. In both wings of the area of distribution, much qualitative variation is present, but quantitative variation was greater in the west. Genes of companion weeds were introduced in the domesticated material through cycles of hybridisation - selection / adaptation. Ervoides has yet to be demonstrated as a donor of genes, but perhaps it is the one responsible for some of the original features of the aethiopicae grex.

iii. The present pattern of distribution of wild lentils can be explained by:
- Migration of wild forms, in mixed populations, with the cultivated species.
- Pre-existence of the wild forms in the original area.
The data best fit the first of these two alternatives.

iv. If the previous hypothesis, that wild forms migrated in mixed populations with the cultivated, is true, the pattern could have originated by either, or a combination of the following:
- Different ecological requirements of nigricans, ervoides and orientalis; the first two are adapted to Mediterranean conditions, and the third to the mountainous regions of southwestern Asia. Ervoides could have been the only one adapted to the savannah environment.
- Stocks of lentils used by western farmers came out from the westernmost fringe of the original region of lentils, where nigricans and ervoides had to be more frequent in primitive domesticated materials. Orientalis, at the opposite end, had to be most frequent in the eastermost fringe of the same area, from where the crop spread eastwards. Only ervoides persisted in the first population which reached Ethiopia.

v. No matter which hypothesis is true, nigricans and orientalis are actual companion weeds of culinaris (and perhaps also ervoides) and donors of genes to culinaris.

vi. The cradle of 'lentoid' characters (viz., flattened pods and seeds) seems to be the region between western Turkey and Kurdistan. It could be the place of origin of _Lens_.

It is interesting to note that, in modern agriculture,the farmers buy seeds from commercial producers who carefully homogenise and purify their stocks. The cultivated species are halted in their evolution, unless breeders reproduce the natural cycle of hybridisation with companion weeds and selection. This is also true for lentils. So it is essential to preserve the genetic potential contained in wild lentils and even in related species. We will learn much more about evolution and, at the same time, we will improve our current cultivated material.

But...there are so many things still to be discovered about lentils...!

Acknowledgements

I am very grateful to Drs. Erskine and Sindhu and, through the latter, to Dr. Slinkard, not only for their comments and suggestions on this chapter, but also for the provision of unpublished information.

References

Alefeld, F. 1866. Landwirtschaftliche Flora. Reprint 1966, ed. Otto Koeltz, Koenigstein-Taunus.

Barulina, H. 1930. Lentils of the USSR and other countries. Suppl. 40 to the Bulletin of Applied Botany, Genetics and Plant Breeding, 265-304. Leningrad. (English summary).

de Candolle, A.P. 1882. Origins of cultivated species. Reprint 1967, London: Hafner.

Davis, P.H. and Plitmann, U. 1970. _Lens_. In Flora of Turkey Vol III, Edinburgh: Edinburgh University Press.

Harlan, J.R. and de Wet, J.M.J. 1971. Towards a rational classification of cultivated plants. Taxon 20: 509-517.

Hopf, M. 1971. Vorgeschichtliche Pflanzenreste aus Ostspanien. Sonderdruck aus den Madrider Mitteilungen 12: 101-114.

Ladizinsky, G. 1979 a. The origin of lentil and its wild genepool. Euphytica 28: 179-187.

Ladizinsky, G. 1979 b. The wild genepool of lentil. LENS 6: 24.

Renfrew, J.M. 1969. The archaeological evidence for the domestication of plants: methods and problems.In The Domestication and Exploitation of Plants and Animals, eds P.S. Ucko and G.W. Dimbleby. London: Duckworth.

van Zeist, W. 1976. On macroscopic traces of food plants in southwestern Asia. Philosophical Transactions of the Royal Society, London B 2751: 27-41.

Williams, J.T., Sanchez, A.M.C. and Jackson, M.T. 1974. Studies on lentils and their variation. I. The taxonomy of the species SABRAO Journal 6: 133-145.

Williams, J.T., Sanchez, A.M.C. and Carasco, J.F. 1975. Studies on lentils and their variation II. Protein assessment for breeding programmes and genetic conservation. SABRAO Journal 7: 27-36.

Zohary, D. 1972. The wild progenitor and the place of origin of the cultivated lentil *Lens culinaris*. Economic Botany 26: 326-332.

17. GENETIC RESOURCES OF LENTILS

M. Solh

Faculty of Agricultural Sciences, AUB, Beirut, Lebanon

W. Erskine

Food Legume Improvement Program, ICARDA, Aleppo, Syria

Introduction

The genetic resources of lentils comprise the primitive varieties or landraces of the cultivated species (Lens culinaris Med.) and its wild relatives within the genus Lens Miller.

Selection on the cultivated lentil, over millennia, has resulted in the development of a myriad of landraces adapted to specific locations. The geographical distribution of the variation among landraces is far from random (Witcombe, Chapter 1), being related to the evolutionary history of the crop (Cubero, Chapter 16) and its spread into regions of high ecological diversity. There is also variation within landraces, which are composed of mixtures of largely homozygous genotypes. Some heterozygosity may also be present in landraces, as has been found in barley populations (Jain and Allard, 1960), in spite of the low level of natural cross-pollination in lentils (Wilson and Law, 1972).

Most of the cultivated area of some two million hectares is sown with landraces. However, lentil improvement has been underway for some years, and improved cultivars have now been released in a few countries. These cultivars are now replacing the landraces and causing genetic erosion. The percentage of the lentil area under landraces in some countries is given in Table 26. In Turkey, the cultivars Kislik-Pul 11, Kislik-Yesil 21, and Kislik-Kirmizi 51 are now grown on 30 to 40% of the lentil area. In Egypt, the cultivar Giza 9 has been released. Improved cultivars such as Pusa-1, Pusa-4, Pusa-6, L-9-12, T-6, T-36, Pant L-406 and Pant L-639 have been developed in India, where they are now causing the genetic erosion of the landraces which they are replacing.

Climatic factors have been causing genetic erosion in Ethiopia, where six years of drought in the north have resulted in the loss of many landraces of lentils. Changes in land use, for various economic reasons, are also causing the genetic erosion of lentils. In southern Europe, lentils have been largely replaced by other crops. Over the 50 year period 1927-1977, the lentil area in Italy declined from 34000 to 2000 ha, partly due to Orobanche infestation and partly for economic reasons (Barulina, 1930; Anonymous, 1981). In the USSR, the area dropped from 42500 to 6000 ha over a similar period. Elsewhere, in India,

J.R. Witcombe and W. Erskine (eds.) Genetic Resources and Their Exploitation - Chickpeas, Faba beans and Lentils. ISBN-13:978-94-009-6133-3 (PB)
© 1984, Martinus Nijhoff/Dr W. Junk Publishers for ICARDA and IBPGR.

Lebanon and Turkey, lentils are also being ousted by crops which are producing a higher economic return.   Some coastal areas of Chile are going out of lentil production because of the yield losses due to rust (Uromyces fabae).   In Jordan and Syria, the cost of labour for the manual harvesting of lentils is compelling some farmers to look for replacement crops. Syria exports a large quantity of lentils, and marked fluctuations in the price of lentils sometimes cause a temporary drop in the lentil area. Any upgrading of rainfed land through the provision of irrigation will also result in the replacement of lentils by more economic crops. All of these changes in land use are resulting in the genetic erosion of lentil landraces.

Table 26. The lentil area of some countries together with percentage of the area sown with landraces and macrosperma cultivars, and the composition of national lentil collections. (Source: ICARDA lentil genetic resources survey 1979).

Country	Lentil area (1000 ha)	% Lentil area under land races	% Lentil area with macrosperma	Number of accessions in national collection	Number of indigenous accessions
Algeria	18.0 (1975)	0	95	49	49
Bangladesh	75.0	70	0	0	0
Canada	16.0	92	98	600	0
Chile	50.0	85	100	150	80
Ethiopia	56.0-180.0	100	0	1600*	100
Greece	4.3	10	25	181	32
India	800.0	-	0	3300	1300
Iraq	9.7 (1978)	100	0(rare)	18	18
Jordan	20.0	100	0	0	0
Lebanon	2.0	100	30	0	0
Spain	70.0	100	100	0	0
Sudan	0.1	0	100	40	0
Syria	178.0 (1977)	100	11	5424	205
Turkey	240.0	60-70	10-15	1000	1000
USA	65.0	0	100	1150	0

* Mostly from the ILL Collection at ICARDA

Unfortunately, very little is known of the extent of genetic erosion of the wild species of lentil. However, it is probable that some of their habitats are being destroyed through overgrazing and other changes in land use.

Concurrent with the recent increase in lentil breeding has been the awareness amongst breeders of the need for a wide genetic base for future improvement. This situation has its parallels in many other crops (see Frankel and Hawkes, 1975). It is now well established that any strategy for the exploitation of a crop must have provision for the collection and maintenance of both landraces and their wild relatives.

Geographic distribution of variability of the cultivated lentil

The distribution of the cultivated lentil in the Old World in 1930 is shown in Figure 24. This distribution has changed slightly, over the past 50 years, with the spread of macrosperma cultivars from Turkey south into Lebanon and Syria, and the large reduction in lentil area in southern Europe. To the east, lentil is important in Bangladesh and Burma, as shown by the areas under cultivation in Table 27. In the New World, lentils are grown on more than 15000 ha in each of Argentina, Canada, Chile, Colombia and the USA.

Within this distribution, some areas show particular diversity. A comprehensive survey of the geographical distribution of variability in lentils was completed by Helena Barulina in 1930. She said that the greatest accumulation of large seeded forms is found in the Mediterranean countries: Spain, Italy with islands, and Greece. As to the small seeded, there are several regions in which their varieties are concentrated: south-western Asia (Afghanistan, Iran), Transcaucasia, western Asia (Asia Minor, Syria and Jordan) and parts of Spain.

Barulina (1928) also reported considerable variability in lentils from Afghanistan. Interestingly, a germplasm collecting expedition to Afghanistan by ALAD (Solh, Rashid and Hawtin, 1974) did not find the range of variability described by Barulina, although a total of 75 samples was collected from nine provinces. It is not clear whether the low level of variability found in 1974 was due to an actual loss of variability since Barulina's survey, or due to incomplete coverage in the recent collection. Nevertheless, for political reasons, fresh collections cannot currently be made. It is clear that collections need to be made promptly to avoid possible future political complications that make the genetic resources unavailable.

Chile has the longest history of lentil cultivation in the New World, where the plant was introduced by the Spanish after 1500 A.D. The crop is grown in the dry coastal areas of Chile and also in the foothills of the Andes, covering a wide range of agro-ecological conditions. It is probable that considerable diversity in lentils exists in these areas because selection for local adaptation, coupled with low gene-flow between adjacent landraces, will have acted to preserve variability among them. There is little variability in lentils in Canada and the USA. Chilean cultivars are grown in Argentina, and the cultivation of lentils in Mexico is also recent. Measurements of the variability of germplasm accessions from Mexico, have confirmed the expectation of low diversity of the Mexican material in comparison with accessions from the Old World.

Information on the geographical distribution of variation in the cultivated lentil can be used to assess how comprehensive the existing germplasm collections really are.

Germplasm collections of the cultivated lentil

Known world germplasm collections in lentils are maintained for breeding purposes in India, the USA and Syria. The Indian Agricultural Research Institute (IARI) in New Delhi maintains more than 3300 accessions from the lentil growing countries of Asia, Africa, Europe and America. WRPIS, at Pullman, Washington, USA, maintains a collection of about 1150 accessions.

ICARDA maintains a collection (ILL) of 5424 accessions from 53 different countries (Table 27). This ILL collection includes almost half of the WRPIS and IARI collections. Other major lentil collections duplicated within the ICARDA collection are shown in Table 28.

Known national lentil germplasm collections are listed in Table 26. The national collections of India and Turkey include a large number of accessions which have originated in their respective countries. There is little recent information about the lentil germplasm collections in East European countries, China and the USSR. Barulina, in 1928 and 1930, reported that 1500 samples from various parts of the world were maintained at the Institute of Applied Botany, Leningrad, USSR.

The geographical distribution and collections of wild lentils

The distributions of the four species of wild lentil are shown in Figure 23. The ease of crossing both Lens orientalis and L. nigricans with the cultivated lentil highlights the importance of the genetic resources of the wild species to the current exploitation of the crop. Some measurements of the variability within the wild species have been made (Williams et al., 1974) but the geographical distribution of the variability within the wild species is completely unknown.

There is no collection containing all four wild species. ICARDA maintains seven accessions of wild lentils, most of which are Lens orientalis. Three species, L. orientalis, L. nigricans and L. ervoides are maintained in the Crop

Table 27. Lentil area in various countries of the world (Anon., 1981) and the number of accessions in the ILL collection from these countries.

Country	Lentil area (1000 ha) in 1981	Number of accessions
AFRICA	119	509
1. Algeria	16	14
2. Egypt	6	85
3. Ethiopia	59	375
4. Libya		1
5. Morocco	34	22
6. Somalia		2
7. Sudan		1
8. Tunisia	4	9
N. & C. AMERICA	79	49
1. Canada		2
2. Costa Rica		1
3. Guatemala		2
4. Mexico	10	24
5. U.S.A.	69	20
S. AMERICA	89	353
1. Argentina	22	6
2. Chile	48	335
3. Colombia	17	8
4. Ecuador	1	
5. Peru	2	3
6. Uruguay		1
ASIA	1562	3969
1. Afghanistan		124
2. Bangladesh	84	36
3. Burma	3	
4. Cyprus		9
5. India	1000	1905
6. Iran	38	902
7. Iraq	10	22

Table 27 contd.

Country	Lentil area (1000 ha) in 1981	Number of accessions
8. Japan		1
9. Jordan	9	294
10. Lebanon	4	70
11. Nepal		12
12. Pakistan	87	37
13. Syria	127	208
14. Turkey	200	313
15. Yemen		37
EUROPE	95	278
1. Albania		2
2. Austria		1
3. Belgium		2
4. Bulgaria	1	23
5. Czechoslovakia	2	17
6. France	12	7
7. Germany (D.D.R.)		22
8. Germany (D.F.R.)		3
9. Greece	4	92
10. Hungary	1	23
11. Italy	2	8
12. The Netherlands		1
13. Norway		1
14. Poland		4
15. Portugal		4
16. Romania		1
17. Spain	73	151
18. U.K.		1
19. Yugoslavia	1	25
USSR	9	103
Unknown	-	52
World	1953	5424

Table 28. Major lentil germplasm collections in the ILL collection.

Number of accessions	Source
75	ALAD Afghanistan Food Legume Collection
102	ALAD field collection in Iraq, Jordan, Lebanon and Syria.
52	Egyptian Collection, Agricultural Res. Station, Giza, Egypt.
323	Ethiopian National Collection, Addis Ababa University, Debre Zeit, Ethiopia.
51	FAO - USDA Collection, Iran.
78	Indian Agricultural Research Institute (IARI), New Delhi, India.
108	Inst. Nacional de Investigaciones Agronomicas (INIA), Cordoba, Spain.
35	Inst. Nacional de Investigaciones Agronomicas (INIA), Madrid, Spain.
54	Inst. de Investigaciones Agropecuarias (INIA), La Platina, Chile.
68	Inst. Nacional de Technologia Agropecuaria (INTA), San Pedro, Argentina.
32	IBPGR collecting mission, Yemen Arab Republic.
851	Pantnagar collection, Pantnagar, India
58	Plant Introduction Centre (PIC), Menemen, Izmir, Turkey.
823	Punjab Agricultural University, Ludhiana, India.
80	Pulses and Oilseeds Research Station, Berhampore, West Bengal, India.
712	Regional Pulse Improvement Program (RPIP), USDA, Karaj, Iran.
96	Research Centre for Agrobotany (NIAVT), Tapioszele, Hungary.
583	WRPIS, Pullman, Washington, U.S.A.
61	Zentral Institut fur Genetik und Kulturpflanzenforschung, Gatersleben, D.D.R.

Development Centre, University of Saskatchewan, Canada. Other universities are known to be studying these species (Ladizinsky, 1979 a and b). Nothing is known of the availability of Lens montbretii.

It is clear that the existing collections of all four wild species are totally inadequate, and a high priority must be given to expanding them. Each species should be collected from as many areas as possible in order to sample the geographical variability within these species. Lens orientalis, L. ervoides and L. montbretii generally flower during April and May, and L. nigricans usually flowers in May (van der Maesen, 1979). A start to the collection of the wild species may be made at those locations, mentioned in local Flora, at which herbarium specimens have been collected.

Collection of lentil germplasm

The gaps in lentil germplasm collections can be identified by relating the geographical distribution of variability to the composition, by countries, of the known collections. On this basis, priority areas for collection have been identified as:

i. Wild species

Lens orientalis, L. nigricans, L. ervoides and L. montbretii. Collection is needed of all four species across their distribution ranges.

ii. Cultivated lentils

Priority areas for the collection of cultivated lentils are shown in Table 29.

Many of the accessions in the ILL collection, the largest of the known world collections, have been collected from market places and along main roads, without any particular sampling strategy in mind. However, important variability is likely to be found in the more inaccessible areas where there is considerable diversity of agro-ecological conditions. National collectors are probably in a better position to collect in such areas. Once a target area for collection has been identified, decisions should be made on the strategy of sampling. In the absence of prior information on the distribution of variability within the target area, systematic sampling on a 'coarse grid' (Hawkes, 1976), and every 10 km along major routes (Witcombe, Chapter 1) have been recommended.

Some accessions in the ILL collection are based on very few seeds, and consequently are not an adequate sample of the variation present within landraces. At every field collection site, pods should be collected from about 40 random plants in order to provide a suitable sample of the population

variability, and also to allow for subdivision of each accession without seed
multiplication.

Table 29. Priority areas for collection of cultivated
lentils.

Continent	Country	Region and comments
Africa	Algeria	
	Morocco	
Asia	Bangladesh	
	Burma	
	India	Bihar and Madhya Pradesh
	Iraq	North and North East
	Pakistan	
South America	Chile	Andean Foothills

Many accessions in the ILL collection are sadly lacking in documentation on
the site of collection; they are only known by the country of collection. For
example, the only information on collection available on 599 accessions from
Iran is that they were collected somewhere in that country. Proper
documentation at the time of collection is vital (Witcombe, Chapter 1). Seed
exchanged by institutes should be accompanied by its collection data.

Maintenance of lentil germplasm

After collection, accessions should be subdivided between the national
germplasm collection and a gene bank for long term storage. The duplication of
accessions will lower the probability of their loss. At ICARDA, some, but not
all, of the accessions are duplicated in other collections. National germplasm
collections are important national resources, and are a prime source of adapted
genetic material for local breeding programs (Haddad, Chapter 8).

The maintenance of the genetic structure of accessions during multiplication or regeneration may be easier in national, rather than in international, germp̄asm collections; genetic changes, due to natural selection, will be lower if the environment of multiplication is similar to that at the original collection site.

At present, 1500 plants of each accession in the ILL collection are grown in plots of six rows, 0.25 m apart and 5 m long, using an augmented experimental design (Federer, 1956). When an accession arrives at a national or international genetic resources centre, it should be grown for both multiplication of seed and preliminary evaluation, whilst taking every precaution to maintain the genetic structure of the accession.

One large multiplication cycle is preferable to frequent cycles of regeneration, to avoid the effects of artificial or natural selection, genetic drift, and the mechanical mixing of seed. Subsamples of the seed from the multiplication should then be used for further evaluation, distribution and long term storage. Locally recommended management practices should be followed during the multiplication. Those characters which are studied in the preliminary evaluation should correspond to those in the descriptor list (Appendix 11) to ensure uniformity in the way the data are collected. In turn, this will facilitate the exchange of information among breeders. The IBPGR will publish a complete list of minimum descriptors and descriptor states for lentil.

The ILL germplasm collection is currently stored in plastic bottles holding around 750 g of seed each, which is maintained at ambient temperature and humidity. Naphthalene crystals are added to the bottles to control weevils. Long term facilities for seed storage are planned for ICARDA, where the lentil collection is currently used as an active collection for breeding purposes.

Variability in some characters of the cultivated lentil

The initial visual impression of low variability in a world lentil collection, compared with other crops, is misleading and attributable to the generally short stature of the crop. A start has already been made on the systematic evaluation of lentil genetic variability for current breeding aims such as resistance to pathogens, e.g. wilt (Fusarium oxysporum f. sp. lentis) (Kannaiyan and Nene, 1976, Khare and Joshi, 1974, Khare and Sharma, 1969), rust (Uromyces fabae) (Agrawal et al., 1976; Nene et al., 1975), powdery mildew (Erysiphe polygoni) (Khare et al., 1971; Mishra, 1973) and pea seed-borne mosaic virus (Muehlbauer, 1977). Germplasm has also been screened for tolerance to drought and salinity (Bhuktiar, 1979, Jana and Slinkard, 1979), seed characteristics (Sharma and Kant, 1975) and resistance to thrips (Hudson et al.,

1973), and pod borer (Etiella zinckenella) (Kooner et al., 1978). Variability within germplasm collections has been found for all of these characters.

Plant breeders, who wish to exploit the variation within germplasm collections, need to know the range of variability available. The range in a number of quantitative characters of importance is shown in Table 30. These ranges were recorded on the ILL collection grown at Tel Hadya, Syria, during the 1978-79 season when the effective precipitation was 235 mm with a fair distribution.

Table 30. Range and mean for six characters measured on the ILL germplasm collection' together with the number of accessions tested and the percentage coefficient of variation (CV) of a systematically repeated check.

Character	Range	Mean	No. of accessions	CV
100-seed weight (g)	1.1-8.6	3.2	3974	-
Crude protein (%)	20.6-33.4	28.1	1863	6.4
Time to maturity (days from sowing)	154-197	170.3	2958	1.6
Plant height (cm)	10-45	25.5	2895	8.0
Height of lowest pod (cm)	6-30	14.1	1772	1.0
Pod number per peduncle	1.0-1.7	1.1	403	2.1

The range in 100-seed weight, recorded on 3974 accessions, was 1.07 to 8.55 g with an overall mean of 3.20 g. This range is greater than those reported by Barulina (1930) and Sharma and Kant (1975). Both accessions with a seed weight less than 1.5 g/100 seeds came from Afghanistan, whereas both accessions with seed weights of more than 8.5 g/100 seeds originated in Syria. The distribution of the accessions for 100 seed weight is continuous (Figure 26). This confirms the observation made by both Williams et al. (1974) and Sharma and Kant (1975) of the arbitrary nature of the division of lentils into the macrosperma and microsperma groups on the basis of seed size.

At ICARDA, for practical reasons, a seed weight of 4.5 g/100 seeds, in lieu of a measurement of seed diameter, has been arbitrarily chosen to separate the microsperma from the macrosperma accessions. The seed diameter is strongly correlated with seed weight (Eser, 1970) and there are no significant differences between the average seed thickness of large and small seeded lentils (Barulina, 1930; Eser, 1970). In the lower latitudes of the Old World (Afghanistan, Bangladesh, Egypt, Ethiopia, India, Pakistan and Sudan),

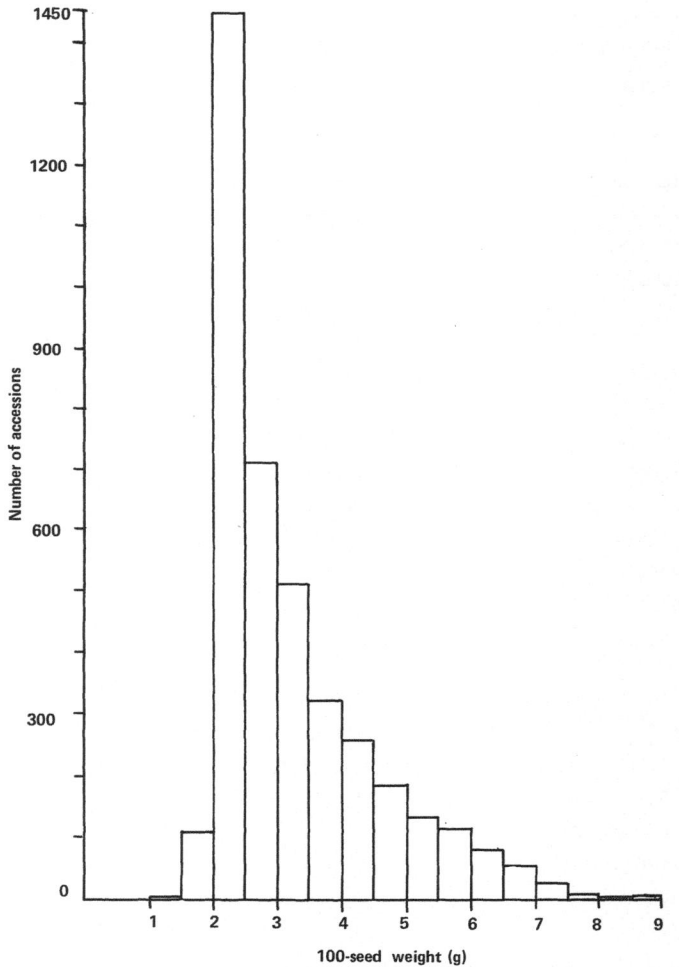

Figure 26. Seed size distribution in 3974 accessions from the ILL collection.

exclusively small seeded accessions are found, and these are earlier maturing than the macrosperma types.

The percentage crude protein (N x 6.25) of 1863 germplasm accessions was measured by a Technicon Auto-Analyser and expressed on a dry weight basis. The range was from 20.6 to 33.4%, with a mean of 28.1%. The range in crude protein reported by Williams et al. (1975) was from 15 to 30%. In the ILL collection, the coefficient of variation of a systematically repeated check variety was only 6.4%, showing the low level of environmental heterogeneity for crude protein in the trial.

The five accessions with the highest protein were from Hungary and Turkey, whereas the five lowest protein entries originated in Ethiopia and Iran. These 10 entries were all in the microsperma group. A particularly wide range in crude protein was found amongst the Ethiopian accessions. The mean protein levels of accessions from Greece, Hungary, Iran, Lebanon, Turkey and the USSR were higher than average. Greek and Turkish material was also found to be high in protein by Williams et al. (1975). Since the correlation between seed weight and crude protein was non-significant, it is possible that high protein macrosperma cultivars could be produced.

The time to maturity, measured on 2958 accessions, ranged from 154 to 197 days. The earliest material was predominantly from Ethiopia and India, but there were also early accessions from both Iraq and Yemen. The coefficient of variation of the systematic check was low (1.6%).

The accessions from the lower latitudes in the Old World were quick to mature, whereas the material from Europe, USSR and Turkey was generally later than average. Interestingly, the accessions from India showed strikingly little variability in time to maturity, and yet, under Indian conditions, considerable variability in both time to flower and maturity has been observed amongst Indian accessions (Kant and Sharma, 1975; ICARDA lentil genetic resources survey). Time to maturity is an example of a character for which screening in one environment does not give an adequate discrimination among the accessions; it requires screening in a wide range of environments.

The range in plant height of the ILL collection was from 10 to 45 cm. The tallest accession was the Canadian-registered variety, Laird (ILL 4349). Generally, accessions from Egypt, Greece and Turkey were taller than average, and the Egyptian material was both early and tall. Tall plants are easier to harvest by hand or machine than short plants.

The mechanical harvesting of lentils will be greatly facilitated by the introduction of cultivars which are both tall and have their pods borne high off the ground. This latter characteristic is particularly crucial in the Near East where lentils are often grown on very stony land. The range in height of the lowest pod was 6 to 30 cm in 1772 accessions. The tall variety Laird also had pods highest-borne above the soil surface.

The average number of pods per pod-bearing peduncle varied from 1.0 to 1.7 over 403 accessions. The mean of the accessions was 1.1 pods per peduncle, highlighting the predominance of single-podded peduncles. The coefficient of variation of the check cultivar was 2.1%. Nine of the ten accessions with the highest number of pods per peduncle, and also nine of the ten accessions with the lowest pod number per peduncle, were microsperma types.

This systematic evaluation of germplasm for current breeding aims illustrates the immediate exploitation of lentil collections. Looking to the future, changes in the pattern of use and cultivation of lentils will alter the aims of improvement programs. The evaluation and exploitation of germplasm will continue directing the evolution of the crop into new dimensions.

Conclusion

A wide range of genetic diversity already exists within collections of lentils. However, these germplasm collections are far from complete and the variability present within the wild lentil species is still largely unexplored. Priority areas for future collection have been identified by considering the geographic distribution of variation in the species and the composition of known germplasm collections.

These gaps should be filled because of the current widespread genetic erosion of lentils. IBPGR and ICARDA are jointly attempting to stimulate the necessary collection and the future maintenance of lentil genetic resources by national programs.

Acknowledgements

The authors wish to acknowledge the contribution of the following people to the 1979 ICARDA lentil genetic resources survey; Mr. Hamed Balehcene, Algeria; Dr. Anwarul Quader Shaikh, Bangladesh; Dr. A.E. Slinkard, Canada; Ing. Juan Tay, Chile; Mr.Geletu Bejiga, Ethiopia; Dr. E. Stylopoulos, Greece; Dr. B. Sharma, India; Directorate of Field Crops, Ministry of Agriculture, Iraq; Dr. Nasri Haddad, Jordan; Dr. Michel Abi Antoun, Lebanon; Dr. J.I. Cubero, Spain; Dr. Farouk A. Salih, Sudan; Dr. D. Eser and Mr. Engin Kinaci, Turkey; and Dr. F.J. Muehlbauer, USA.

References

Agrawal, S.C., Khare, M.N. and Agrawal, P.S. 1976. Field screening of lentil lines for resistance to rust. Indian Phytopathology 29: 208.

Anonymous. 1981. FAO Production Yearbook. Rome, Italy: FAO.

Barulina, H. 1928. Lentils of Afghanistan. Bulletin of Applied Botany, Genetics and Plant Breeding, Leningrad. (English summary).

Barulina, H. 1930. Lentils of the USSR and other countries. Pages 265-304 in Suppl. 40 to the Bulletin of Applied Botany, Genetics and Plant Breeding, Leningrad. (English summary).

Bhuktiar, B.A. 1979. Lentil germination under simulated moisture stress and salinity. M.Sc. Thesis, American University of Beirut, Beirut, Lebanon.

Eser, D. 1970. Turkiye'de Yetistinlen mercimek cesterlinin onemli morfolojik karakterlin uzerinde arastirmalar. Ziraat Fakultesi Yayinlan: 383, Bilimsa, Arastimave Incelemeler: 233, Ankara University. (English summary).

Federer, W.T. 1956. Augmented (or Hoonnuiaku) designs. Hawaiian Planters Records 55: 191-208.

Frankel, O.H. and Hawkes, J.G. (Eds.) 1975. Crop Genetic Resources for Today and Tomorrow. Cambridge, U.K.: Cambridge University Press.

Hawkes, J.G. 1976. Manual for field collectors (seed crops). AGPE: Miscellaneous 7. Rome, Italy: FAO.

Hudson, L.W., Dietz, S.M., Davis A.M. and Pesho G.R. 1973. Lentil inventory Lens esculenta: Catalog of seed available from the Western Regional Plant Introduction Station. Washington State University, Washington, USA: USDA.

Jain, S.K. and Allard, R.W. 1960. Population studies in the predominantly self-fertilized species. I. Evidence for heterozygote advantage in a closed population of barley. Proceedings National Academy of Science, U.S. 46: 1373-1377.

Jana, S.K. and Slinkard, A.E. 1979. Screening for salt tolerance in lentils. LENS 6: 25-27.

Kannaiyan, J. and Nene, Y.L. 1976. Reactions of lentil germplasm and cultivars against three root pathogens. Indian Journal of Agricultural Science 46: 165-167.

Kant, K, and Sharma, B. 1975. Variation in flowering time of lentil under Indian conditions. LENS 2: 15-16.

Khare, M.N. and Joshi, L.K. 1974. Studies on wilt of lentil. Annual Report 1973-74 of PL-480 scheme, J.N. Agricultural University, Jabalpur, India.

Khare, M.N. and Sharma, H.C. 1969. Field screening of lentil varieties against Fusarium wilt. Mysore Journal of Agricultural Science 4: 354-356.

Khare, M.N., Mishra, R.P., Joshi, L.K. and Sharma, H.C. 1971. Diseases of rabi pulses in Madhya Pradesh - screening of varieties for disease resistance. in Proceedings, All India Workshop on Rabi Pulses, Indian Council for Agricultural Research, New Delhi.

Kooner, B.S., Singh, B. and Singh, K.B. 1978. Preliminary screening of lentil germplasm against pod borer, Etiella zinckenella Triet. LENS 5: 1-3.

Ladizinsky, G. 1979 a. The wild genepool of lentil. LENS 6: 24.

Ladizinsky, G. 1979 b. The origin of lentil and its wild genepool. Euphytica 28: 179-187.

Mishra, R.P. 1973. Studies on powdery mildew of lentil (Erisyphe polygoni BC) in Madhya Pradesh. Punjabras Krishi Vidyapeeth Research Journal 2: 72-73.

Muehlbauer, F.J. 1977. Resistance in lentils to pea seed borne mosaic virus. LENS 4: 31.

Nene, Y.L., Kannaiyan, J. and Saxena, G.C. 1975. Note on performance of lentil varieties and germplasm cultures against Uromyces fabae (Pers.) de Bary. Indian Journal of Agricultural Science 45: 177-178.

Sharma, B. and Kant, K. 1975. Variability for seed characters in the world germplasm of lentil. LENS 2: 12-14.

Solh, M., Rashid, K. and Hawtin, G.C. 1974. Food legume collection. Afghanistan July-August 1974. Arid Lands Agricultural Development Program, Lebanon: Ford Foundation. (mimeographed).

van der Maesen, L.J.G. 1979. Genetic resources of grain legumes in the Middle East. Pages 140-146 in Food Legume Improvement and Development, eds. G.C.

Hawtin and G.J. Chancellor: IDRC-ICARDA.

Williams, J.T., Sanchez, A.M.C. and Jackson, M.T. 1974. Studies on lentils and their variation. I. The taxonomy of the species. SABRAO Journal 6: 133-146.

Williams, J.T., Sanchez, A.M.C. and Carasco, J.F. 1975. Studies on lentils and their variation. II. Protein assessment for breeding programs and genetic conservation. SABRAO Journal 7: 27-36.

Wilson, V.E. and Law, A.G. 1972. Natural Crossing in Lens esculenta Moench. Journal of the American Society of Horticultural Science 97: 142-143.

Apperdix 11. A list of lentil descriptors.

PASSFORT DATA

1. ACCESSION DATA   (As per IBPGR standard format)

2. COLLECTION DATA   (As per IBPGR standard format)

CHARACTERIZATION AND PRELIMINARY EVALUATION DATA

3. SITE DATA   (As per IBPGR standard format)

4. PLANT DATA

4.1 VEGETATIVE
        4.1.1 Seedling pigmentation
        4.1.2 Leaf pubescence
        4.1.3 Leaflet size
        4.1.4 Plant height
        4.1.5 Tendril length
4.2 INFLORESCENCE AND FRUIT
        4.2.1 Time to flowering
        4.2.2 Time to maturity
        4.2.3 Flower ground colour (standard)
        4.2.4 Number of flowers per peduncle
        4.2.5 Pod pigmentation
        4.2.6 Number of seeds per pod
4.3 SEED
        4.3.1 100 seed weight
        4.3.2 Ground colour of testa
        4.3.3 Pattern on testa
        4.3.4 Colour of pattern on testa
        4.3.5 Cotyledon colour

FURTHER CHARACTERIZATION AND EVALUATION

5. SITE DATA   (As per IBPGR standard format)

6. PLANT DATA
    6.1 VEGETATIVE
        6.1.1 Lodging susceptibility

      6.1.2 Biological yield

6.2 INFLORESCENCE AND FRUIT

      6.2.1 Height of lowest pod

      6.2.2 Pod shedding

      6.3.3 Pod dehiscence

6.3 SEED

      6.3.1 Seed yield

      6.3.2 Protein content

      6.3.3 Methionine and other sulphur amino acids

      6.3.4 Cooking time

7. STRESS SUSCEPTIBILITY

    7.1 Low temperature

      7.1.1 Winter kill

      7.1.2 Low temperature damage

    7.2 Drought

    7.3 High soil moisture

    7.4 Salinity

8. PEST AND DISEASE SUSCEPTIBILITY

    8.1 Pests

      8.1.1 *Aphis craccivora* cowpea aphid

      8.1.2 *Sitona* spp. weevil

      8.1.3 *Bruchus* spp. weevil

      8.1.4 *Etiella zinckenella* pod borer

      8.1.5 Other (specify in the NOTES descriptor, II)

    8.2 Fungi

      8.2.1 *Uromyces fabae* rust

      8.2.2 *Ascochyta* spp. blight

      8.2.3 *Fusarium oxysporum* f.sp. *lentis* vascular wilt

      8.2.4 *Peronospora lentis* downy mildew

      8.2.5 Other (specify in the NOTES descriptor, II)

    8.3 Bacteria

    8.4 Virus

    8.5 Parasitic weeds

      8.5.1 *Orobanche* spp.

      8.5.2 Other (specify in the NOTES descriptor, II)

9. ALLOENZYME COMPOSITION

10. CYTOLOGICAL CHARACTERS AND IDENTIFIED GENES

11. NOTES

18. EVALUATION AND UTILIZATION OF LENTIL GERMPLASM IN AN INTERNATIONAL BREEDING PROGRAM

W. Erskine

Food Legume Improvement Program, ICARDA, Aleppo, Syria

Introduction

The lentil (<u>Lens culinaris</u> Med.) is an annual, self-pollinating, pulse crop which is commonly grown around the Mediterranean and in parts of south and west Asia. Its area of cultivation around the world amounted to 1.84 million hectares in 1980, at an average yield of 655 kg/ha (Anonymous, 1981). India is by far the largest producer (Figure 27).

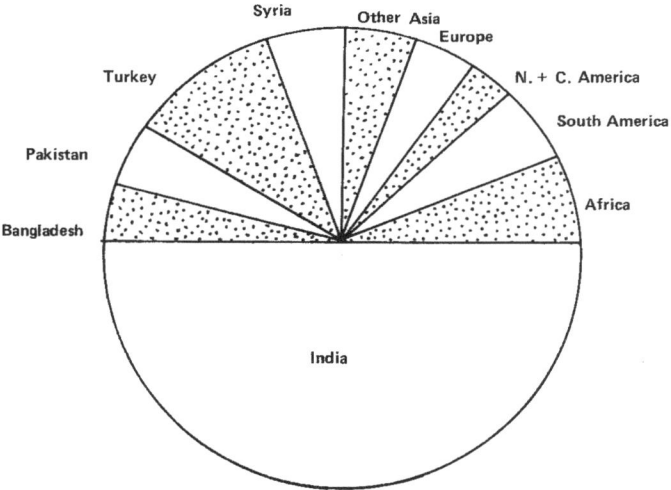

Figure 27. World lentil production in 1980.

Little research has been done on the crop, so it is under-exploited. There are few improved cultivars available to farmers and these are concentrated in India. The landraces of lentil have low yield potential, specific adaptation, and are often unresponsive to improvements in husbandry such as early sowing and phosphate application. Consequently, the major objective in lentil breeding is to increase yields. Lentils are susceptible to some insect pests, diseases, and

J.R. Witcombe and W. Erskine (eds.) Genetic Resources and Their Exploitation - Chickpeas, Faba beans and Lentils. ISBN-13:978-94-009-6133-3 (PB)
©1984, Martinus Nijhoff/Dr W. Junk Publishers for ICARDA and IBPGR.

paras˙tic weeds. In the Mediterranean region, _Sitona_ spp. of weevil and the paras˙tic weed _Orobanche_ _crenata_ are the major reducers of yield, whereas in south Asia and Ethiopia, rust (_Uromyces_ _fabae_) and root rot/wilt complex are the most important. Many of the existing landraces have a plant growth habit that is di˙ficult to harvest mechanically, and many farmers have stopped cultivating the crop because of the prohibitively high cost of hand harvesting.

Breeding aims

The main breeding aims are to develop cultivars or genetic stocks with high and stable seed yields for each of the three major agro-ecological regions of lenti˙ production. Cooking and nutritional quality, and nitrogen-fixing ability should be maintained or improved. The additional specific traits required for the regions are as follows:

i.   High altitude region: Cold tolerance for winter planting and attributes to facilitate harvesting (tall, erect, non-lodging growth habit; high pod retention and pod indehiscence).

ii. Middle to low elevation areas in the Mediterranean region: High biological yield; tolerance to _Orobanche_ spp. and _Sitona_ spp. of weevil; resistance to root rot/wilt complex; attributes to facilitate harvesting; tolerance of droughty conditions.

iii. Region of more southern latitudes (including Egypt, Ethiopia, India, Pakistan and Sudan, etc.): Earliness through reduced sensitivity to photoperiod and temperature; resistance to root rot/wilt complex and rust.

Direct exploitation of germplasm

Table 31. Seed yield of selected entries from the lentil regional small-seeded yield trial (1980-81) at various sites.

Selection Number	Accession Number (ILL)	Seed Yield (kg/ha)			
		Syria	Lebanon	Jordan	Mean
78S 26003	8	1643	1141	1724	1458
76TA 66088	223	1670	933	2004	1441
Jordan local	4354	1572	841	1697	1304
Lebanon local	4399	1011	918	1150	1001
Syria local	4400	1354	971	1393	1209
Standard error	+	236	416	332	

After characterization and preliminary evaluation for morpho-agronomic characters, the variability within the germplasm collection can be exploited. Selections were made both within and between accessions. For example, in the case of grain yield, single plants were selected from within accessions, examined in progeny rows, and finally tested extensively in replicated yield trials. A response to selection for grain yield has been achieved from this approach in Syria, Lebanon and Jordan (Table 31).

Selection for specific characters

Following evaluation, not only has yield been selected for, but also other specific characters of economic importance.

Cold tolerance

Lentils in the high plateau areas of Afghanistan, Iran and Turkey are currently planted in the spring. However, research in Turkey revealed that there were genotypes which could tolerate severe winter conditions (e.g. ILL 1878). Earlier planting in winter gives a considerable yield advantage over spring sowing if the cultivars are suitably cold tolerant. A study was, therefore, undertaken to identify other cold tolerant lentil accessions from a world germplasm collection.

In collaborative work between ICARDA and the Turkish Ministry of Agriculture, 3592 accessions from the ICARDA world lentil collection were screened at Hymana (altitude 1055 m; 39° 50' N, 32° 40' E), Ankara, Turkey (Erskine, Meyveci, and Izgin, 1981). They were planted in mid-October, 1979, in unreplicated 2 m rows with two local checks repeated after every ten accessions. Plant emergence was scored prior to winter. The plants were covered with snow for 47 consecutive days; the lowest air temperatures, recorded in December 1979, and in January, February and March 1980, were -9.0^oC, -26.8^oC, -12.3^oC, -13.8^oC, respectively. After the snow melted, the proportion of plants to survive undamaged was scored.

The two checks, ILL 1878 and 1880, were undamaged and considered cold tolerant. A total of 238 accessions, comprising 6.6% of the 3592 accessions screened, were cold tolerant (Table 32). From the Asian material, there were 47 cold tolerant accessions from Iran, 35 from Turkey, and 29 from Syria. Amongst the European material there were 15 cold tolerant accessions from Greece. The largest number (58) of cold tolerant accessions came from Chile. At least 45 accessions were screened from all these countries. A lower number of accessions from Iraq, Jordan and Tunisia were examined, and five cold tolerant accessions were found from each country. From the results of screening, it is clear that the probability of finding additional material with cold tolerance is highest

Table 32. The origin of germplasm accessions showing cold tolerance in Turkey.

Country	Number of cold tolerant accessions screened	Total number of accessions screened	Percentage of cold tolerant accessions
ASIA			
Afghanistan	2	106	2
India	13	1358	1
Iran	47	884	5
Iraq	5	19	26
Jordan	6	32	17
Lebanon	3	67	5
Nepal	0	8	0
Pakistan	0	22	0
Syria	29	88	33
Turkey	35	240	15
USSR	2	38	5
Yemen	0	4	0
AFRICA			
Algeria	1	8	13
Egypt	0	56	0
Ethiopia	0	246	0
Morocco	2	10	20
Tunisia	5	8	63
EUROPE			
Czechoslovakia	0	2	0
Cyprus	0	6	0
France	0	1	0
Germany	0	4	0
Greece	15	46	33
Hungary	1	27	4
Italy	2	5	40
Spain	3	9	33
UK	1	1	100
Yugoslavia	2	18	11
AMERICAS			
Argentina	0	2	0
Chile	58	208	28
Colombia	1	5	20
Costa Rica	0	1	0
Ecuador	1	2	50
Guatemala	1	1	100
Mexico	1	24	4
Peru	0	1	0
USA	0	1	0
Australia	1	2	50
UNKNOWN	1	32	3
TOTAL	238	3592	7

amongst further accessions from the following seven countries: Chile, Greece, Iraq, Jordan, Syria, Tunisia and Turkey. In these areas there has been natural selection for cold tolerance under cultivation.

The cold tolerant germplasm comprised 68% large seeded accessions (100-seed weight > 4.5 g), whereas the ICARDA world collection has only 15% large seeded accessions overall. The predominance of cold tolerant material with large seeds is because such cultivars are commonly grown in those countries where natural selection for cold tolerance has occurred. There has been no natural selection for cold tolerance in countries where lentils rarely experience cold conditions For example, there were no cold tolerant accessions from either Egypt or Ethiopia, and less than two per cent of the accessions from Afghanistan, India and Pakistan were tolerant.

Tolerance to Orobanche spp.

Parasitic weeds of the genus Orobanche are a major problem in lentil production in parts of west Asia and north Africa. Once an area is infected with Orobanche seed, chemical treatment and plant husbandry cannot control infestation. The use of resistant or tolerant cultivars would alleviate the problem. Since genetic resistance to Orobanche spp. was unknown in lentils, screening was undertaken at ICARDA.

During the 1979-80 season, 1000 germplasm accessions were grown at Tel Hadya in four replications in soil heavily infected with Orobanche crenata. There were highly significant differences between the entries screened for susceptibility to Orobanche. Sixty two accessions had a maximum of two emerged Orobanche inflorescences/m^2. The most susceptible entry was ILL 1386 from Iran which supported an average of 46 Orobanche inflorescences/m^2. It is now being used as a susceptible control.

Table 33. Average number of inflorescences/m^2 of Orobanche crenata on selected lentil accessions in the 1980-81 season.

Accession number	Number of inflorescences/m^2
ILL 3047	1.8
ILL 3112	2.2
ILL 4400 (Syrian local large)	19.8
ILL 4401 (Syrian local small)	16.8
L.S.D. 5%	7.2

In the following season, the most resistant entries were planted in infested soil in six replicates at Kafr Antoon, where significant differences in susceptibility were again found. Genotypes ILL 3047 and 3112, both from India, were the most resistant to O. crenata; they had an average of only 1.8 and 2.2 Orobanche inflorescences/m^2 in comparison to 16.8 inflorescences/m^2, on the best local check (Table 33). These genotypes have been included in the crossing block to incorporate the tolerance to Orobanche into high yielding cultivars.

Disease resistance

This aspect of lentil improvement was given a relatively low priority over the last two years, but some progress has been made.

Lentil rust (Uromyces fabae) is the major fungal disease affecting lentils in the Indian sub-continent and Ethiopia. A rust nursery comprising rust resistant lines from Pantnagar, India, and local controls was grown at Lattakia in Syria. The Indian resistant lines were all free of rust, showing the similarity of the rust in Syria to that in India.

Genetic differences in susceptibility to Ascochyta lentis were observed in germplasm grown at Tel Hadya. For example, Syrian local large (ILL 4400) was susceptible, whereas Syrian local small (ILL 4401) was resistant.

Downy mildew is becoming important in India (Khare, 1981) and may sometimes cause economic damage under conditions of high rainfall and humidity in the Mediterranean region. Eleven germplasm accessions were identified in Lebanon as resistant to Peronospora lentis, and they have been used in the crossing block.

Improved plant type for mechanical harvesting

Lentil is harvested by hand in the Middle East, where the high cost of harvest labour is driving some farmers from its cultivation. In Canada and the USA lentils are mechanically harvested by cutter-bar. Harvesting by cutter-bar and, in view of the value attached to lentil straw in the Middle East, pulling techniques which involve the collection of the entire above-ground part of the crop, may offer economic alternatives to the traditional method of harvest.

Ninety per cent of the lentils grown in Syria are too short for harvesting by either method. A selection program was therefore undertaken to produce taller cultivars from the locally adapted material.

A landrace population (ILL 4401) was irradiated with y-rays from a cobalt source at 10-20 k rads. The irradiated seeds were grown in Tel Hadya during the 1978-79 season as the M_1 generation. In the following season 12000 seeds from the M_1 were sown. At harvest, single plant selection for increased plant height was undertaken by eye. Random plants were taken from the M_2 population and from the un-irradiated base population. The seed from these populations was thus:

i.   Un-irradiated and unselected.

ii.   Irradiated and unselected.

iii.   Irradiated and selected.

Measurements of average plant height were made on the progenies of these single plants during the 1980-81 season at Tel Hadya.

The average height of the selected population was 31.4 cm, which was significantly greater than the height of the other populations (Table 34). Clearly, eyeball selection for height with an intensity of < 1% gave a response to selection.   The tallest progeny were 38.9 cm high, and they were from the un-irradiated and unselected base population, emphasising the variability within the landrace.   There were no significant differences between the populations in the variance for plant height, and it is evident that irradiation did not increase the variability for height.   It follows that selection could have been practised as effectively on the original un-irradiated landrace.

Table 34. Population size and average height (cm) of three populations of ILL 4401.

Population	Number of selections	Height
Control	48	30.1 ± 0.47
Irradiated	45	29.1 ± 0.40
Irradiated and selected for height	81	31.4 ± 0.30

Tall genotypes have been identified within the germplasm collection.   These tall accessions are being hybridised with locally adapted populations, like ILL 4401, to produce tall, adapted segregants. This strategy offers more scope for increasing the plant height to 35 cm than selection within the indigenous land races. A total of 66 crosses with tall, erect parents were made in the 1980-81 season.

Considerable progress has also been made in selecting tall, erect and high yielding material, which can be harvested mechanically. An erect selection, 78S 26002 (ILL 8) from Jordan, yielded 20 and 23% more than the best check at Tel Hadya in the 1979-80 and 1980-81 seasons, respectively.

Adaptation to southern Asia

There is a discontinuity in the adaptation of lentils in the northern hemisphere of the Old World.   Genotypes originating from western Asia and southern Europe are very late to flower when grown in southern Asia (Bangladesh, India, and Pakistan) at photoperiods of less than 11 h; so much so, that they produce few or no pods because of the onset of adverse environmental conditions during their period of reproductive growth. For example, Syrian local large (ILL 4400), a macrosperma type (100-seed weight > 4.5g), flowered only three days after Giza 9 (ILL 784), from Egypt, when they were grown in northern Syria; whereas the difference in time to flower was 37 days in Pakistan where the photoperiod was less than 11 h (Table 35). This has precluded the use of macrosperma genotypes in south Asia, where larger seeded material is preferred. However, a macrosperma germplasm accession, Precoz (ILL 4605), which was from Argentina and had a seed size of 5.2 g/100 seeds, flowered and reached maturity a fortnight before the local check when grown near Delhi, India. This accession can assist in broadening the genetic base of the crop in south Asia if used either directly, as an introduction, or in hybridisation.

Table 35. Time to flowering in two contrasting environments. (Data from Lentil Adaptation Trial, 1979-80).

Particulars	Locations	
	Islamabad, Pakistan	Tel Hadya, Syria
Latitude (^{o}N)	33	37
Photoperiod (h)*	10.9	12.2
Mean temperature** (^{o}C) Maximum	19	16
Minimum	6	6
Mean time to 50% flowering (d)		
Cultivar - Giza 9 (ILL 784)	108	96
Cultivar - Syrian local (ILL 4400)	145	99

* At 7 d prior to flowering (includes twilight).
** Average of month prior to flowering.

Information on the adaptation of the crop is important in formulating a rational strategy for breeding. Thus, it is clear that selections for south Asia must be made under the appropriate environmental conditions of photoperiod and temperature. Accordingly, crosses are being made at ICARDA between south Asian material and both macrosperma genotypes and selections from Egypt and Ethiopia.

The progeny of these crosses are distributed as yield trials of bulk populations at the F_3 generation. Selections are then made between and within crosses by national programs. In addition, a Lentil International Early Screening Nursery has been initiated for southern Asia.

Exploitation following hybridisation

During the evaluation of the germplasm, many genotypes with important traits were identified and included in the crossing program for recombination with other desirable characters; 112 and 224 bi-parental crosses were made in the 1978-79 and 1979-80 seasons, respectively. There were forty six entries from 18 countries in the crossing block in 1980-81, and these were crossed in 240 two-way combinations and 20 three-way combinations. The aims of these recombinations, together with the target region of each cross, are shown in Table 36. The varied origins of the material in the crossing block reflects the emphasis on establishing a wide genetic base for lentil improvement. This is of particular importance because of the paucity of breeding effort on lentil elsewhere in the region.

Table 36. Crossing block in the 1980-81 season, showing the target region
         and the aims of recombination.

Region	Recombination aims	No. of cross combinations
High Altitude	Cold tolerance x High yield in Turkey;	7
	Attributes for harvest x High yield in Turkey.	18
Middle to low elevation (Mediterranean)	Orobanche diallel;	27
	High yield x 3 pods per peduncle;	23
	High yield (small seeds) x High yield (large seeds);	29
	High yield x Attributes for harvest;	48
	High yield x F1 (locals x attributes for harvest).	18
Southern Latitudes	High yield in Indian sub-continent x High yield in N. and E. Africa	90
		Total 260

Following recombination, a bulk-pedigree system of breeding is being followed. An off-season summer nursery at Shawbak, Jordan, has been used for generation advancement, which allows two cycles per year. Single seed and pod descent methods were tried, but they failed because of the high level of plant attrition in the summer nursery.   Bulk populations of promising crosses are tested at the F_3 generation in international F_3 trials.

The selection of single plants is undertaken in the F_5 generation.   (A schematic summary of the breeding method utilized is given in Figure 28). Single plant progenies are grown in unreplicated rows with systematically repeated controls.   A total of 19941 progeny rows were examined at Tel Hadya over the 1979-80 and 1980-81 seasons (Table 37).   The overall selection pressure was 15.7% in these nurseries.

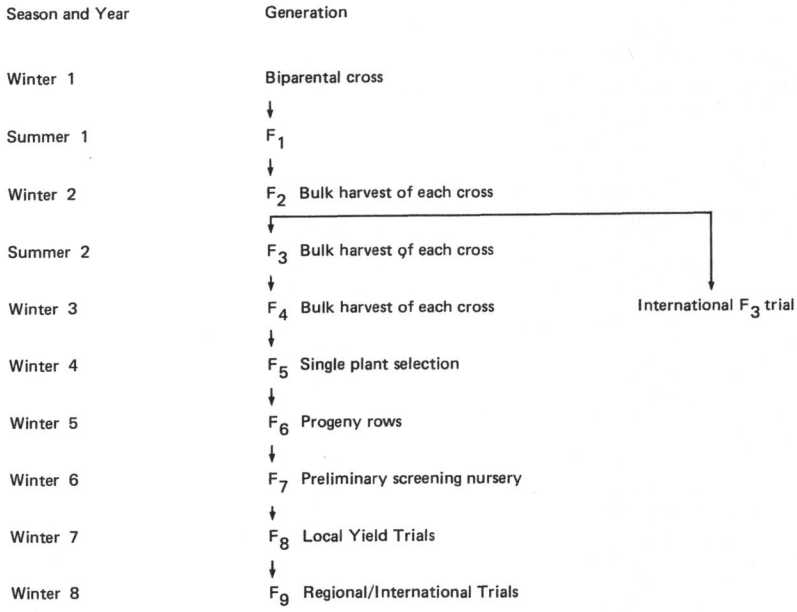

Figure 28.   The scheme for breeding at ICARDA

Table 37. The number of single plant progenies tested over two seasons, together with the selection pressure (%) exercised in these nurseries.

	1979-80		1980-81	
	Large Seeded	Small Seeded	Large Seeded	Small Seeded
Number of test entries	1388	8959	3383	6211
Selection pressure (%)	14.3	19.4	16.5	10.3

The selected progeny rows are individually bulk harvested, and grown as the F_7 generation in preliminary screening nurseries in the following season, in an augmented experimental design. A total of 5586 progenies have been tested, at either Tel Hadya or Terbol, over the seasons 1979-80 and 1980-81. Amongst these entries, 506 selections have exceeded the mean of the best check in seed yield (Table 38).

Table 38. Summary of preliminary screening nurseries grown at Tel Hadya, Syria, and Terbol, Lebanon, in 1979-80 and 1980-81. Syrian local large (ILL 4400) and small (ILL 4401) were checks in all trials.

Location	Season	Seed size	Number of test entries	Number of entries exceeding best check	Check mean seed yield (kg/ha)
Tel Hadya	1979-80	Large	225	14	1216
Tel Hadya	1979-80	Small	1745	102	1264
Tel Hadya	1980-81	Large	199	34	1679
Tel Hadya	1980-81	Small	1738	147	1598
Terbol	1979-80	Large	344	85	1048
Terbol	1979-80	Small	538	108	1071
Terbol	1980-81	Large	304	8	576
Terbol	1980-81	Small	493	8	968
Total			5586	506	

Selections from the preliminary screening nurseries are grown in replicated yield trials in Tel Hadya, and, if seed permits, also in Lebanon. Over the 1979-80 and 1980-81 seasons a total of 814 selections were planted in replicated local yield trials at Tel Hadya, with 410 of these also grown in Lebanon. The results of these trials are summarized in Table 39.

The best entries in these trials are fed into the international screening nurseries, regional yield trials, and into the crossing block.

The requirements of national programs for either macrosperma or microsperma genetic stocks, taken into consideration with the difference in morphology of the taxa, led to a division of the breeding program into large and small seeded streams. Between 1979 and 1981, 80% of the selections tested were small seeded (< 4.5 g/100 seeds), with the remaining 20% entries having larger seeds. The emphasis on the development of small seeded material reflects its importance in production globally.

Table 39. Summary of results for seed yield in local yield trials in Lebanon (Kfardan and Terbol) and Syria (Tel Hadya) in 1979-80 and 1980-81.

Description of trials	No. of test entries	No. of entries exceeding best check*	Yield of best test entry (kg/ha)	Mean coefficient of error variation (%)
Large seeded, 79-80 Tel Hadya	99	24	1704	25.7
Small seeded, 79-80 Tel Hadya	275	81	2302	24.5
Large seeded, 80-81 Tel Hadya	110	11	2320	18.4
Small seeded, 80-81 Tel Hadya	330	34	2183	18.4
Large seeded, 79-80 Terbol	14**	0	1738	17.4
Small seeded, 79-80 Terbol	22	5	1622	18.1
Large seeded, 80-81 Terbol	88	3	1592	26.6
Small seeded, 80-81 Terbol	110	3	1972	23.2
Small seeded, 80-81 Kfardan	176	15	1071	19.8

* Syrian local large (ILL 4400) and small (ILL 4401) were checks in all trials.
** All entries tested in Lebanon were also tested in Syria.

Regional and international yield trials have been separately prepared for elite large and small seeded selections, but have been sown simultaneously and alongside each other. The mean yield of a total of eleven large seeded trials grown in Lebanon and Syria over the 1980 and 1981 seasons was 1091 kg/ha, whereas the mean figure for small seeded trials was 1131 kg/ha. It is thus clear that the yield potential of large and small seeded selections is similar in Syria and Lebanon.

References

Anonymous, 1981. FAO Food Production Year Book, 1981. Rome, Italy: FAO.

Erskine, W., Meyveci, K. and Izgin, N. 1981. Screening a world lentil collection for cold tolerance. LENS 8: 5-9.

Khare, M.N. 1981. Diseases of lentils. Pages 163-172 <u>in</u> Lentils, eds. C. Webb and G. Hawtin, Slough, U.K.:Commonwealth Agricultural Bureaux-ICARDA.

INDEX

A

Abdalla, M.M.F. 91, 92, 146, 157, 165, 168
Accession book 5, 6, 7, 39, 42, 107
Accession data 160, 222
Accession numbers 5, 7, 9, 11, 22, 37, 40, 41, 43, 48, 53, 77, 108, 109, 226
Accessions, duplicate 20, 46, 63, 107, 146, 213
Acetylene reduction technique 52
Active collection 13, 27, 36, 153, 154, 156, 214
Adaptation trial 126
Adaptation (to poor soils) 81, 232
Adapted races 76
Adler, A. 96, 99, 100, 105, 121
Advanced cultivars (see cultivars, improved)
Adzuki bean 74
Aegean Regional Agricultural Research Institute, Izmir Turkey 107
Afghanistan 49, 86, 97, 105, 106, 140, 141, 146, 148, 181, 196-197, 207, 209, 215, 227-229
Afghanistan-Indian-Turkemenian border 197
Africa 76, 96, 106, 163, 208, 209, 213, 225, 228
Agar culture 54, 58, 69
    Malt extract agar 69
    Potato-dextrose agar 69
Agar plate method 69
Agrawal, P.S. 218
Agrawal, S.C. 214, 218
Agricultural Research Centre, Egypt 157
Agricultural Research Institute, Cyprus 146
Agrobacterium 52
Agro-climatic conditions 113
    -ecological conditions 207, 212
    -ecological regions 226
    -economic characteristics 107
    -environments 126
Agronomic traits 89
    Morpho- 124, 174, 227
Air conditioning 24, 27, 30
Air-lock (ante-room) 27
Akkadians 198
ALAD 43, 107, 119, 207, 211
ALAD Afghanistan Food Legume Collection 211
ALAD field collection in Iraq, Jordan, Lebanon & Syria 211
Alaska pea 78
Albania 210
Albizia 14

Aldrin 64
Alefeld, F. 191, 192, 202
Aleppo, Syria 51
Alfalfa mosaic virus 162
Algeria 106, 140, 146, 148, 206, 209, 213, 228
Allard, R.W. 179, 186, 205, 219
Alleles 21, 150, 174
Alloenzyme composition 162, 223
Allogamous species 76, 132, 167
Allogamy 132
Alloploid 79, 81, 91
Alloploidy 79, 81
Alphanumeric 41, 42, 46
Alternaria spp 65, 69, 161
    A. alternata 65
Altitude 9, 40, 43
Ambient conditions (temperature and humidity) 24, 25, 26, 28, 29, 30, 156, 214
Amdoun, Tunisia 127
Amino-acid 173, 223
Amman, Jordan 85
Amphiploid 82
Amphiploid route 82
Analysis of variance 46, 48, 49, 50, 119
Anatolia, Turkey 96, 197
Andes 207, 213
Andigena crosses 78
Aneuploid 91
Anishetty, M. 146, 158
Ankara, Turkey 86
Anthocyanin 99
Anti-nutritional factors 183-184
Aphids 91, 161, 180, 181, 182
Aphis craccivora 223
    A. fabae 181
Aquadulce 181
Arachis spp 79
    A.batizocoli 74
    A.cardenasii 74
    A.hypogea (groundnut/peanut) 51, 74, 78, 79, 81-83
    A.monticola 74
Archeological remains 96, 140, 142, 193
Archeological sites 195
Archeological studies 197
Argentina 87, 148, 207, 208, 209, 228, 232
Arid Lands Agricultural Development Program, Ford Foundation (see ALAD)
Army worms 180
Ascochyta spp 69, 70
    A. blight 66, 69, 70, 87, 90, 91, 113, 114, 116, 121, 125, 128, 161, 165, 180, 181, 182, 223